BIOLOGY
A Self-Teaching Guide

Steven D. Garber

WILEY

A Self-Teaching Guide
John Wiley & Sons, Inc.

New York • Chichester • Brisbane • Toronto • Singapore

Illustration permission credits listed by page number.

11 B. S. Neylan and B. G. Butterfield, *Three-Dimensional Structure of Wood,* Chapman and Hall Ltd, Publishers / **20** Dr. W. H. Wilborn (Figure 2.5a) / **20** Dr. D. E. Kelly (Figure 2.5b) / **21** Dr. D. W. Fawcett / **25** photomicrograph from T. Naguro, K. Tanaka, *Biological Medical Resources Supplement,* 1980, Academic Press, Harcourt Brace Jovanovich, Inc. / **27** Dr. H. J. Arnott / **49** Dr. H. W. Beams (Figure 3.1b) / **52** R. A. Boolootian, Ph.D., Science Software Systems, Inc. / **66** Carolina Biological Supply Company / **122** drawing adapted from Buchsbaum and Pearse, *Animals Without Backbones,* University of Chicago Press, 1987 / **182** Huxley M. R. C., Cambridge, England / **208** W. B. Saunders Company / **209** drawing adapted from A. C. Guyton, *Textbook of Medical Physiology, 6th edition,* W. B. Saunders Company / **217** Dr. Thomas C. Hayes / **220** Abbott Laboratories, Chicago, IL / **262** Springer-Verlag Publishers Inc., New York / **263** from *Mosaic* (May/June 1975) National Science Foundation / **285** Dr. F. A. Eiserling / **286** Dr. K. Amako / **306, 307, 324** drawings by Alice Baldwin Addicott / **308** drawing adapted from Gibson, *Edible Mushrooms and Toadstools,* Harper & Row Publishers, Inc., 1899 / Drawings appearing on pages **341, 344, 345, 346, 347, 348** by Jerome Lo / All line drawings not listed above drawn by John Wiley & Sons Illustration Department.

Library of Congress Cataloging-in-Publication Data

Garber, Steven D.
  Biology : a self-teaching guide.

  (Wiley self-teaching guides)
  Bibliography : p.
  1. Biology—Study and teaching. 2. Biology.
I. Title.   II. Series.
QH316.G37 1989   574'.076   89-5506
ISBN 0-471-62581-7

Printed in the United States of America
89 90 10 9 8 7

To Dad

Whose advice and love
were enough to last a lifetime

How to Use This Book

The objective of this Self-Teaching Guide is to take what is currently being taught in most college biology courses, and make it *easier to learn*. And in so doing, the student is more successful. This book organizes, condenses, and clarifies the main concepts and terms, highlighting all the primary points needed to fully grasp the material and to do well in any biology course.

The chapters and questions provide excellent preparation for quizzes, tests, exams, prelims, and finals. This Self-Teaching Guide has a study section at the end of each chapter where key terms, questions to think about, and multiple-choice questions like those that appear on exams (complete with the answers), are presented. Plus, this book is the perfect study companion when preparing for standardized exams in biology such as the Student Aptitude Test (SAT), The American College Testing Program Assessment [ACT], Admissions Testing Program Achievement Test, in Biology, Advanced Placement Program: Biology, College Level Examination Program: Subject Examination in Sciences—Biology, National Teacher Examinations [NTE] Specialty Area Test: Biology and General Sciences, and the Graduate Record Examinations (GRE): Subject Test—Biology.

Preface

Having taught biology at several colleges and universities, as well as having taught at the high school level, has given me some special opportunities. For one, I have been on all sides of the fence: I've been the student, the TA, and the professor. I have worked with students at all levels, and we've used many different biology texts. The need is clear, the similarities are apparent. Students almost unanimously agree, biology courses need something. They force you to learn too much too fast. And what you're taught and what you read is usually too hard to understand.

Having remedied biology courses' weaknesses and built on their strengths, this book allows us to collaborate, bridging the gap between what's being taught and what's being learned. Although I'd prefer to be there teaching you all personally, through *Biology: A Self-Teaching Guide*, the net effect will be positive. The formula works!

Acknowledgments

It was David Sobel, my editor, mentor, and friend who relayed the sales force's message that students need this book! I thank you all, including the marketing department and the buyers at the bookstores who hear requests and recognize the student's needs, and encouraged all of us to remedy the situation. Efficient learning improves students' lives, not to mention their GPAs, improving study habits, attention spans, information retention, and overall capacity. Promoting biological literacy is a noble task. And I'm sure I speak for my compatriots when I say wherever we clarified the occasional conundrum and its ancillary ambiguities, it was and is our pleasure.

Only through perseverence and persistence did we achieve our goal. With lots of caring and hard work—by such talents as Stephen Kippur, Ruth Greif, Frank Grazioli, Corinne McCormick, Nana Clark, Margie Schustack, all of whom represent the John Wiley and Sons, Inc. brain trust, were we able to create a top-quality final product. In addition, Alison Shurtz and Cristina Escobar of Publication Services added many of the finishing touches, for which I am grateful.

I would also like to thank Joanna Burger, my Ph.D. advisor in the Graduate Program in Ecology at Rutgers University whose good example I hope to emulate. And, I thank my many colleagues at Yale University, Cornell University, The University of Kansas, New York University, The American Museum of Natural History, The City University of New York, The National Park Service, and Rutgers University with whom I have had countless valuable interactions that have made it possible to, in my own small way, contribute to what we like to call, higher education.

The love and/or support from my parents (Mom and Herb) and siblings (step and blood) is particularly appreciated. As is that of friends, including Julie Cooper who was my mole in the Mayor's office, and Betsy Lerner, without whose symplesio- and synapomorphies, I don't know what I would have done. In recompense to Sue Kaufman, how about a hand shake, a kiss on the cheek, a beer, and/or a walk in the woods? Special combinations will be considered on an ad hoc basis.

My literary agent, F. Joseph Spieler, has guided and directed while graciously allowing me to clear the way for upcoming projects. Many thanks to the editors for whom I write articles and *The Urban Naturalist* columns; your understanding when multiple obligations demanded more time than we had was like a breath of fresh air, something one hardly gets anymore.

Contents

CHAPTER 1

Origin of Life

The evolution of life on Earth has involved the following sequence of events. The first living things to appear were the simplest creatures, one-celled organisms. From these came more complex, multicellular organisms. Becoming more complex meant more than just an increase in cell number. With more cells came cellular specialization, where certain cells within the multicellular organism carried out specific tasks. Millions, even billions of years of organismal changes led to the living things we now call plants and animals.

Since this basic sequence of events is in accord with that agreed upon by most geologists, paleontologists, biologists, and even theologians, one might conclude that Moses, Aristotle, and Darwin were all keen observers and naturalists who were able to logicially assess the most probable creation story.

Scientists generally concur that the time from the formation of our solar system until now has been on the order of some 4.5 billion years. Those who believe the world as we know it was created in six days are often called **creationists**. Their method of inquiry is based on the belief that the Bible is to be accepted as a completely accurate accounting of all about which it speaks. **Scientists**, on the other hand, utilize what they call the **scientific method**, which allows them to test hypotheses and theories, and to develop concepts and ideas. However, there are many good scientists who also happen to be creationists. Even though the two are often compared and contrasted, the fact is that creationism is not a science, and therefore it is not dealt with in most biology books.

SPONTANEOUS GENERATION

An early hypothesis concerning the origin of living organisms from nonliving material is known as **spontaneous generation**. This concept had many adherents for over a thousand years. Aristotle believed insects and frogs were generated from moist soil. Other elaborations on this basic theme prevailed for centuries. It wasn't until 1668 that Francesco Redi, an Italian, challenged the concept of spontaneous generation when he tested the widespread belief that maggots were generated from rotting meat. He placed dead animals in a series of jars, some of which were covered with a fine muslin that kept flies out while allowing air in. Other jars containing dead animals were left open. Maggots appeared only on the meat in the jars that were left open. In these, flies had been able to lay their eggs, which then hatched into fly larvae, or maggots. The flies were unable to land on the meat in the covered jars, and no maggots appeared there. From this he concluded that maggots would arise only where flies could lay their eggs.

During the nineteenth century, following other experiments, the theory of spontaneous generation of microorganisms was laid to rest by experiments conducted in France by Louis Pasteur and in England by John Tyndall. They demonstrated that bacteria are present in the air, and if the air surrounding a heat-sterilized nutrient broth is bacteria-free, then the broth remains bacteria-free. Until this time, people still believed microorganisms arose spontaneously.

One last vestige of mysticism in the debate concerning spontaneous generation had to be invalidated before theories regarding the origin of life could move ahead; this was known as the **vitalist doctrine**. Adherents of this idea maintained that life processes were not determined solely by the laws of the physical universe, but also partly by some **vital force**, or **vital principle**. By the late 1870s, most scientists agreed that all organisms arose from the reproduction of preexisting organisms, and the concept of spontaneous generation had become history.

CONDITIONS FOR THE ORIGIN OF LIFE

Life is thought to have developed here on earth through a sequence of chemical reactions over time. The most widely held hypothesis begins with the formation of the sun and the planets, which coalesced from a cloud of matter that resulted from a supernova, an old star that had exploded. Given the same explosion and the same amount of time, the same sequence of events would probably happen again, though the results might not be quite the same.

In what became our solar system, the largest mass to coalesce became our sun, and one of the smaller masses became our earth. On earth, the heavier materials sank to the core of the planet while the lighter substances are now more concentrated at the surface. Among these are hydrogen, oxygen, and carbon—important components for all life that eventually evolved.

The primordial atmosphere on earth was considerably different from that which currently exists. The present atmospheric gases are composed primarily of molecular nitrogen (N_2, 78%) and molecular oxygen (O_2, 21%), with a small amount of carbon dioxide (CO_2, 0.033%) and many other gases, such as helium and neon, found in only trace amounts.

The composition of today's atmosphere differs markedly from that found here when life was just beginning to evolve. At that time, the atmosphere contained far more hydrogen, and unlike now, there was very little oxygen. In such an atmosphere, the nitrogen probably combined with hydrogen, forming ammonia (NH_3); the oxygen was probably found combined with hydrogen in the form of water vapor (H_2O), and the carbon occurred primarily as methane (CH_4). The moderately high temperatures of the earth's crust continually evaporated any water that rained into the form of water vapor. As the earth cooled, rain water washed dissolved minerals into low areas creating lakes, seas, and oceans. In addition, volcanic activity erupting in the oceans and on land brought other minerals to the earth's surface, many of which eventually accumulated in the oceans, such as the various types of salts. It should also be mentioned that long before there was any life on earth, the seas contained large amounts of the simple organic compound methane.

Most of the compounds necessary for the development of the initial stages of life are thought to have existed in these early seas. Other studies have indicated that suitable environments for the first steps leading to living material could have existed elsewhere as well. But these environments are still poorly understood, and their potential connection with the origin of life is unclear.

EXPERIMENTAL SEARCH FOR LIFE'S BEGINNINGS

In the 1920s, S. I. Oparin, a Soviet scientist, investigated how life could have evolved from the inorganic compounds that occurred on earth billions of years ago. His work is credited with leading to important later advances, most prominent of which were Stanley Miller's experiments during the 1950s. Miller duplicated the chemical conditions of the early oceans and atmosphere and provided an energy source, in the form of electric sparks, to generate chemical reactions. He found that when warm water and gases containing the compounds presumed to be found in the early oceans and in the earth's primordial atmosphere were subjected to sparks for about a week, organic compounds were formed.

Subsequent experiments, such as those performed by Melvin Calvin and Sydney Fox, have shown that many of the important so-called building blocks of life, or the amino acids that make up proteins, form quite readily under circumstances similar to those first established experimentally by Stanley Miller.

The thin film of water found on the microscopic particles that make clay has been shown to possess the proper conditions for important chemical reactions. Clays serve as a support and as a catalyst for the diversity of organic molecules involved in what we define as living processes. Ever since J. Desmond Bernal presented (during the late 1940s) his ideas concerning the importance of clays to the origin of life, additional prebiotic scenarios involving clay have been proposed. Clays store energy, transform it, and release it in the form of chemical energy that can operate chemical reactions. Clays also have the capacity to act as buffers and even as templates. A. G. Cairns-Smith analyzed the microscopic crystals of various metals that grew in association with clays and found that they had continually repeating growth patterns. He suggested that this could have been related to the original templates on which certain molecules reproduced themselves. Cairns-Smith and A. Weiss both suggest clays might have been the first templates for self-replicating systems.

Some researchers believe that through the mutation and selection of such simple molecular systems, the clay acting as template may eventually have been replaced by other molecules. And in time, instead of merely encoding information for a rote transcription of a molecule, some templates may have been able to encode stored information that would transcribe specific molecules under certain circumstances.

Other scenarios have been suggested to explain how the molecules that make more molecules could have become enclosed in cell-like containments. Sydney Fox and coworkers found that molecular boundaries between protein-nucleic acid systems can arise spontaneously. They heated amino acids under dry conditions and found that long polypeptide chains were produced. These polypeptides were then placed in hot water solutions, and upon cooling them, the researchers found that the polypeptides coalesced into small spheres. Within these spherical membranes, or **microspheres**, certain substances were trapped. Also, lipids from the surrounding solution became incorporated into the membranes, creating a protein-lipid membrane.

Oparin said "the path followed by nature from the original systems of protobionts to the most primitive bacteria . . . was not in the least shorter or simpler than the path from the amoeba to man." His point was that although the explanations intended to show how organic molecules could have been manufactured in primitive seas or on clays seem quite simple, and although one can see how such molecules could have been enclosed inside lipid-protein membranes, taking these experimental situations and actually creating living cells is a tremendous leap that may have taken, at the very least, hundreds of millions of years, perhaps considerably longer.

PANSPERMIA

Although most modern theorists do not accept the idea that living organisms are generated spontaneously, at least not under present conditions, most do believe that life could have and probably did arise spontaneously from non-living matter under conditions that prevailed long ago, as described above. However, other hypotheses have also been suggested for the origin of life on earth.

In 1821 the Frenchman Sales-Guyon de Montlivault described how seeds from the moon accounted for the earliest life to occur on earth. During the 1860s, a German, H. E. Richter, proposed the possibility that germs carried from one part of the universe aboard meteorites eventually settled on earth. However, it was subsequently found that meteoric transport could be discounted as a reasonable possibility for the transport of living matter because interstellar space is quite cold (-220°C) and would kill most forms of microbial life known to exist. And even if something had survived on a meteor, reentry through the earth's atmosphere would probably burn any survivors to a crisp.

To counter these arguments, in 1905 a Swedish chemist, Svante Arrhenius, proposed a comprehensive theory known as **panspermia**. He suggested that the actual space travelers were the spores of bacteria that could survive the long periods of cold temperatures, and instead of traveling on meteors that burned when plummeting through the atmosphere, these spores moved alone, floating through interstellar space, pushed by the physical pressure of starlight.

The main problem with this theory, overlooked by Arrhenius, is that ultraviolet light would kill bacterial spores long before they ever had a chance to reach our planet's atmosphere. This explains the next modification to the theory.

Francis Crick, who along with James Watson received the Nobel Price for discovering the structure of DNA, coauthored an article with Leslie Orgel, a biochemist, in 1973. Their article, "Directed Panspermia," was followed by the book *Life Itself*, in which Crick suggests that microorganisms, due to their compact durability, may have been packaged and sent along on a spaceship with the intention of infecting other distant planets. The only link missing from Crick's hypothesis was a motive.

PROBING SPACE FOR CLUES OF LIFE'S ORIGINS ON EARTH

Recent information concerning the origin of life has opened new avenues of research. To the surprise of many, spacecraft that flew past Halley's Comet in 1986 sent back information showing the comet was composed of far more organic matter than expected. From that, and additional evidence, some have concluded that the universe is awash with the chemical precursors of life. Lynn Griffiths, chief of the life sciences division of the National Aeronautics and Space Administration, said "everywhere we look, we find biologically important processes and substances."

We have known for years, from fossil evidence, that bacteria appeared on earth about 3.5 billion years ago, a little more than 1 billion years after the solar system formed. The great challenge has been to learn how, within that first billion years, simple organic chemicals evolved into more complex ones, then into proteins, genetic material, and living, reproducing cells.

As this current theory stands, it is felt that some 4 billion years ago, following the formation of the solar system, vast quantities of elements essential to life, including such complex organic molecules as amino acids, were showered onto earth and other planets by comets, meteorites, and interstellar dust. Now seen as the almost inevitable outcome of **chemical evolution**, these organic chemicals evolved into more complex molecules, then into proteins, genetic material, and living, reproducing cells.

Unfortunately, no traces of earth's chemical evolution during the critical first billion years survive, having all been obliterated during the subsequent 3.5 billion years. Biologists and chemists now feel, however, that clues concerning the first stages in the origin of life on Earth can be found by looking elsewhere in the solar system. Planetary scientists are to be launching new probes that will eventually investigate these questions, looking for evidence revealing the paths of chemical evolution that may have occurred, or may still be occurring, on planets, moons, comets, and asteroids.

Key Terms

chemical evolution

creationist

microspheres

panspermia

scientific method

scientist

spontaneous generation

vital force

vital principle

vitalist doctrine

Chapter 1 Self-Test

QUESTIONS TO THINK ABOUT

1. Briefly discuss the major theories concerning the origin of life. Give their strong points and their weak points.
2. What is the role that clay is theorized by some to have played in the origin of life?
3. Researchers have experimentally searched for life's beginnings by duplicating the chemical conditions of the early oceans and atmosphere in the lab. Describe some of their results and the implications they hold for the origin of life.
4. Discuss some of the proposed explanations for the origin of life on Earth that suggest life came here from another place.
5. What recent clues to life's origins on Earth have come from space probes?

MULTIPLE-CHOICE QUESTIONS

1. People who believe the Biblical explanation that the world and all its creatures were created in six days are known as:

   a. evolutionary biologists
   b. molecular biologists
   c. systematists
   d. cladists
   e. creationists

2. Scientists use what they call _____ , which allows them to test hypotheses and theories, and to develop concepts and ideas.

   a. Occam's razor
   b. religious dogma
   c. religious faith
   d. scientific method
   e. creation science

3. Aristotle believed insects and frogs were generated from nonliving components in moist soil. This early hypothesis concerning the origin of living organisms is know as _____ .

    a. evolution
    b. spontaneous generation
    c. materialism
    d. creationism
    e. Aristotelian generation

4. Adherents of the _____ maintained that life processes were not solely determined by the laws of the physical universe, but rather, they also depend on some vital force, or vital principle.

    a. dogmatic principle
    b. Darwinian approach
    c. vitalist doctrine
    d. Lamarckian principle
    e. all of the above

5. The composition of today's atmosphere differs markedly from that found here when life was just beginning to evolve. At that time the atmosphere contained far more _____ .

    a. hydrogen
    b. oxygen
    c. potassium
    d. iridium
    e. all of the above

6. When the chemical conditions of the early oceans and atmosphere are duplicated in the lab and provided with an energy source in the form of electric sparks, _____ (has) have been formed.

    a. life
    b. organic molecules
    c. amino acids
    d. a and b
    e. b and c

7. _____ (has) have been shown to serve as a support and as a catalyst for the diversity of organic molecules involved in what we define as living processes.

    a. quartz crystals
    b. gold
    c. plutonium
    d. clay
    e. all of the above

8. When researchers heated amino acids under dry conditions, long polypeptide chains were produced. When these chains were placed in a hot water
solution and then allowed to cool, the polypeptides coalesced into small
spheres called _____ , within which certain substances were
trapped. Molecules that make more molecules could have become enclosed
in such cell-like containments.

    a. cells
    b. cell membranes
    c. cell walls
    d. microspheres
    e. all of the above

9. It was proposed that germs would have been carried to earth from another
part of the universe via meteorites. Such transport was finally discounted,
however, because:

    a. heat generated during entry into the earth's atmosphere would burn any
       germs to a crisp
    b. no such life was ever found on meteorites
    c. nothing could possibly survive interstellar space
    d. all of the above
    e. none of the above

10. _____ , the comprehensive theory proposed in 1905 by the
Swedish chemist Svante Arrhenius, stated that spores of bacteria that
could survive the long periods of cold travelled alone through interstellar
space, pushed along by the physical pressure of starlight.

    a. panspermia
    b. Arrheniusism
    c. microspermia
    d. germspermia
    e. intergalactic sporesia

ANSWERS

1. e	4. c	7. d	10. a
2. d	5. a	8. d	
3. b	6. e	9. a	

CHAPTER 2

Cell Structure

MICROSCOPES

Anything as small as a cell was unknown before sophisticated optics became available. During the seventeenth century, ground lenses that were being used for eyeglasses were first arranged at opposite ends of a tube, creating a small telescope. It was a short step to the invention of the **microscope**; one of the first was constructed by the Dutchman **Antonie van Leeuwenhoek**.

With this **light microscope**, the examination of specimens was facilitated by thinly slicing them, allowing light to pass through. By staining the specimens, it was possible to emphasize internal structures. For instance, staining cellular fluids pink and staining solid, hard structures purple provided increased contrast, enabling those who study cells, known as **cytologists**, to discern these structures more clearly.

While some light microscopes permit researchers to view objects at as much as 1500 times (1500×) their actual size, **stereomicroscopes**, also called **dissecting microscopes**, magnify objects from only 4× to 80×. With two eyepieces, the advantage to this low-powered, three-dimensional view is that researchers can investigate much larger objects, such as the venation of insect wings (see Figure 2.1).

Since the invention of the light microscope, the most significant technological advance for cell researchers was the development of the **electron microscope** (EM), which occurred in the early 1930s. It not only improved the ability to see smaller structures with greater **magnification**—so they appeared larger—but also enhanced the ability to see things more clearly, or with added **resolution**.

When it was discovered that the illumination of specimens with blue light under the light microscope lent considerably greater resolution than with any colors of longer wavelengths, researchers speculated that using shorter wavelengths might add even more resolution. However, wavelengths shorter than those of violet light are not visible to the human eye. This problem led to the invention of the **transmission electron microscope** (TEM), which utilizes a beam of electrons that travel in shorter wavelengths than those of photons in visible light. These electrons are passed through a thinly sliced specimen within a vacuum to prevent any electrons from being deflected and absorbed

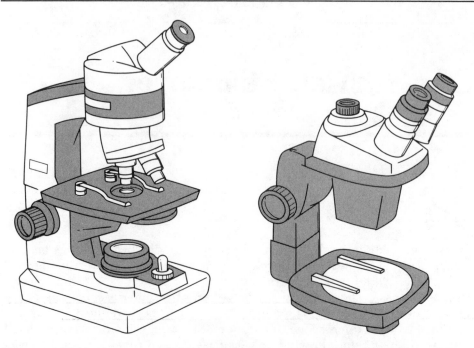

Figure 2.1 *Light microscrope (left) and stereomicroscope (right). The resolving power of the light microscope rarely exceeds a magnification of 1500×. The stereomicroscope, sometimes called a dissecting microscope, has two eyepieces which render relatively large objects three-dimensional. Magnification ranges from 4× to 80×.*

by the gas molecules in the air. Then the electrons are focused with electromagnets on a photographic plate, producing an image that is considerably better than that obtained with a light microscope.

Then in the 1950s the **scanning electron microscope** (SEM) was invented, which focuses electrons that bounce off the specimen. Since the SEM has less resolving power than the TEM, it doesn't require a vacuum and allows researchers to view some smaller organisms alive.

The most recent advances in microscope technology allow scientists to observe living cells with even greater magnification. One of the newly developed techniques is called **contact X-ray microscopy**. Another new microscope, the **scanning tunneling microscope**, has enabled scientists to photograph molecules!

CELLS

With the availability of the first microscopes, researchers began to observe the microscopic structure of many substances, and in 1665 the Englishman **Robert Hooke** described having seen what he called **cells** in a piece of cork. He used this term because the cork appeared to be composed of thousands of

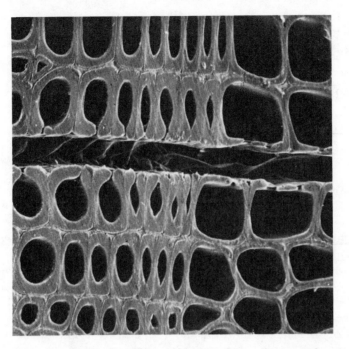

Figure 2.2 *Linearly arranged cell walls in the annual growth rings of a pine tree. In the aggregate, it is the cellulose of each cell wall that gives the tree its rigidity.*

tiny chambers that resembled the individual sleeping rooms in monasteries at the time, which were called cells. He was not aware that he was viewing just the cell walls, which were the only structures remaining from what had once been living cells.

Hooke's initial discovery led to other advances, such as the finding that unlike plant cells, which have thick **cell walls**, animal cells lack such a wall and instead have only a thinner, generally more flexible **plasma membrane** (see Figure 2.2).

Cells were then found to exist independently or as one small part of an organism consisting of many cells, a **multicellular organism**. Hooke was the first to discover that some organisms consist entirely of a single cell. These **unicellular organisms**, such as thousands of species of bacteria and protozoa, carry out all necessary life-supporting functions within one cell without the help of other cells. In contrast, multicellular organisms have cells with specific functions, and together the aggregate of cells embodies a complex organism.

CELL THEORY

It took about 150 years after Hooke discovered cells before several important related facts were articulated. Two German scientists, **Matthias Schleiden** and **Theodor Schwann** were the first to explain, in 1838 and 1839 respectively, the basic tenets of what we now call **cell theory**:

1. Cells are the fundamental units of life.
2. Cells are the smallest entities that can be called living.
3. All organisms are made up of one or more cells.

CELL STRUCTURE AND CELL SIZE

The longest cells are certain nerve cells (neurons), which can reach over a meter in length. While an ostrich egg is 1,500 times the size of a human egg cell–which is 14 times the size of a human red blood cell, itself as much as 35 times the size of many small single-celled microorganisms–most cells do have one thing in common: They tend to be quite small. While the size range reflects considerable diversity, most cells are 0.5 to 40 microns in diameter (1,000 microns equals one millimeter).

Small cell size is thought to be a function of the restriction placed on them by the ratio of surface area to volume. Cells are constantly absorbing molecules from the surrounding medium and releasing molecules into the surrounding medium. These processes are more readily accomplished when a cell is small and the ratio of surface area to volume is quite large. As a cell increases in size, the amount of volume inside the cell increases much more rapidly than the amount of surface surrounding the cell, and in time the cell becomes too large to maintain a stable internal environment.

Many scientists believe that it is more difficult for the nuclear material to maintain control over the entire internal environment when a cell is over a certain size. Therefore, if a small nucleus is most often the rule, then an upper limit is placed on the size of most cells.

CYTOPLASM AND NUCLEOPLASM

Except for the nucleus (or nuclei), everything within the plasma membrane is called the **cytoplasm**. The **nucleoplasm** consists of the contents within the nuclear membrane.

The cell's interior is composed largely of a complex **solution** as well as a heterogeneous **colloid**.

A *solution* is a homogeneous mixture of two or more components in which the particles of the different substances are so small that they cannot be distinguished.

A *colloid* usually contains particles that are too small to be seen but are large enough not to form a true solution. The particles don't settle out at an appreciable rate. Different areas inside a cell may be in different colloidal states.

Some areas are in a **sol** state–that is, the colloidal particles, which are usually macromolecules, are randomly dispersed throughout the area.

The other parts of a cell may be in a **gel** state, in which the colloidal particles interact and form a spongy network. A colloid in the gel state forms a semisolid.

Changes from the sol to the gel state may be stimulated by changes in *pressure, agitation, temperature, pH, salt concentration, or the concentration of other substances.*

Some mixtures containing proteins and lipids have properties of true solutions as well as some sol-gel capabilities of a colloid, and this often typifies the internal fluid environment of a cell.

CELL MEMBRANE

The **cell membrane** is the membrane surrounding the cytoplasm at the cell's surface. These membranes contain varying amounts of *proteins* and *lipids*. The specific types of lipids, which are fat-soluble substances, located in the cell membrane are primarily *phospholipids* and *steroids* (steroids are discussed in more detail in Chapter 14, on Nutrition). The total thickness of the *cell membrane*, which is sometimes also called the *plasma membrane*, is only about 80 angstroms (there are 10 million angstroms per millimeter).

MOVEMENT THROUGH THE CELL MEMBRANE

In addition to providing protection, shape, and strength to the cell, the membrane helps regulate the flow of materials in and out of the cell. Many processes work in concert to maintain the cell's constant internal environment. The condition of constant internal environment is referred to as **chemical homeostasis**, and the methods a cell employs to maintain this constant environment are described below. The two basic methods require different energy input. When the movement of molecules occurs on its own, without any organized energy input from the cell, the movement is termed **passive transport**. When the cell expends energy to move molecules from one location to another, the process is known as **active transport**.

Diffusion

One type of passive transport, called **diffusion, may occur in a dynamic system such as a gas or liquid, in which particles may be moving in a manner that appears random** (see Figure 2.3). In a system where diffusion occurs, a net movement of particles can be observed over time from regions of higher concentration of that substance to regions of lower concentration of that substance.

The particles moving from the area of greater concentration will produce a **concentration gradient** ranging from the point of highest concentration to

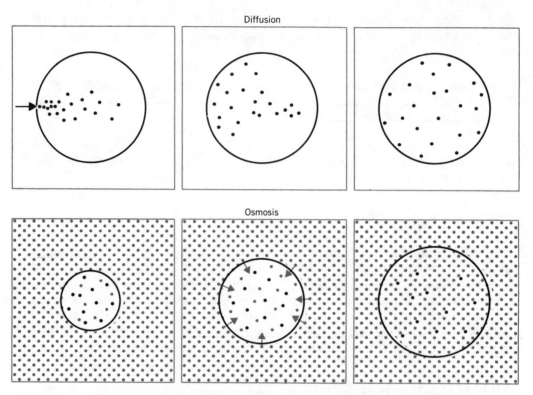

Figure 2.3 *The top three boxes illustrate* **diffusion**. *Dissolved molecules enter the enclosed area and move about within the medium (either liquid or gas). Eventually the molecules become randomly distributed throughout the enclosure. The bottom three boxes illustrate* **osmosis**. *The percentage of dissolved materials inside the enclosure is greater than the percentage outside the enclosure. To counteract the difference, water molecules diffuse through the membrane. The osmotic process continues until the percentage of water molecules inside the membrane equals the percentage outside, unless the membrane becomes stretched to capacity and can no longer accept additional molecules. At such a point it either remains firm (turgid), or it may burst, or in some cases a mechanism such as active transport pumps water out of the cell.*

the point farthest away, where the lowest concentration occurs. The particles don't move in a straight line, but the concentration gradient is instead a result of many, small movements. The tendency is for molecules to move down a concentration gradient–that is, they go from an area of highest concentration of that molecule to where the concentration is less. This movement leads to the point when the molecules are eventually dispersed evenly throughout the environment where they are all able to move equally freely.

The movement of the particles can proceed at different rates, depending on such factors as the concentration and temperature of the medium. Some membranes are impermeable to the movement of anything through them. Other membranes may exclude some substances, but they are permeable to others.

A partially or differentially permeable membrane is called a **semipermeable membrane**.

Osmosis

Another example of passive transport is **osmosis, which involves the movement of water through a semipermeable membrane** (see Figure 2.3). The pressure exerted by dissolved particles in a solution or a colloid, known as **osmotic pressure**, moves water across a semipermeable membrane. When a semipermeable membrane separates two solutions of pure water, the two sides are at **equilibrium**—that is, the osmotic pressure is equal on both sides. When the osmotic pressure is the same on both sides of a semipermeable membrane, each solution is said to be **isosmotic** or **isotonic**. This means that each medium has the same concentration of osmotically active particles, and each has the same osmotic pressure, so neither tends to lose or gain appreciable quantities of water via osmosis.

Another alternative exists when a cell is in a medium with a higher concentration of osmotically active dissolved or colloidal particles, rendering a higher osmotic pressure. In such a situation, water moves across the cell membrane into the medium that is **hyperosmotic** (also called **hypertonic**) relative to the solution inside the cell.

When a cell is in a medium with a lower concentration of osmotically active dissolved or colloidal particles, and the solution thus has a lower osmotic pressure than that inside the cell, the cell is said to be in a medium that is **hypoosmotic** (or **hypotonic**) relative to it, so water moves into the cell.

To avoid confusion, one should be careful to note the point of reference of these terms. For instance, when the internal environment of a cell has a higher concentration of osmotically active particles than the external medium, this means the medium inside the cell is hyperosmotic relative to the external environment. Either way one chooses to state it, the water still moves from the hypoosmotic environment into the hyperosmotic environment. Water moves from where the osmotic pressure is low to where the osmotic pressure is high (to where there are more osmotically active particles). If this process is allowed to continue, all things being equal, the solutions on either side of the semipermeable membrane should eventually become isosmotic to each other.

If particulate matter is poured into one of the two previously isosmotic solutions, and the particles dissolve into solution, then the equilibrium will be upset and the osmotic pressure that now exists between the two sides will be unequal. Water will move from the area of lower concentration of dissolved molecules.

The permeability of different types of plasma membranes varies significantly. An amoeba's plasma membrane has less than one percent of the permeability to water of the membrane of a human red blood cell. Most plasma membranes are quite permeable to water and other substances such as many simple sugars, amino acids, and lipid-soluble substances. Plasma membranes are relatively impermeable to larger complex molecules such as polysaccharides and proteins.

Permeability to small molecules indicates there are pores through which these particles can pass. It may be that these pores open and close according to the cell's needs at a given time. The membrane's permeability to lipid substances is related to the lipids that make up part of the membrane. Lipids are able to dissolve right through the membrane. In the cell membrane are **carrier molecules** that help move certain substances in and out of the cell. They are described in the following discussions on facilitated diffusion and active transport.

Facilitated Diffusion

One of the two main types of transport that involve active, as opposed to passive transport, is known as **facilitated diffusion**. Here, carrier molecules move substances either in or out of a cell in a manner that increases the rate of diffusion across the membrane. The mechanism involved combines molecules that are to be moved across the membrane with the carrier molecules in the membrane. In such a manner, molecules are picked up on one side of the membrane and released on the other side.

A common example of facilitated diffusion is the rate at which cells take up glucose, an important simple sugar in many biological systems. Most cells absorb glucose much more rapidly than the rates at which such absorption would occur without any added help. In concert with the carrier molecules, as the cell starts to use the glucose rapidly, the concentration on the inside of the cell decreases, and because of the increased concentration gradient, glucose is moved into the cell more rapidly than usual.

Facilitated diffusion works in both directions. Sometimes it is equally important to move substances into the cell as it is to move other substances out of the cell. Certain hormones can affect the rates at which some substances move across cell membranes. For instance, insulin increases the rate of glucose uptake by some tissues. While glucose normally moves along a concentration gradient, facilitated diffusion helps increase that rate.

Active Transport

To move substances such as glucose across a cell membrane against a concentration gradient, energy has to be expended. The process by which substances move against their concentration gradients is termed *active*. There are many types of **active transport** systems. Some take up amino acids; others move peptides, nucleosides, or potassium. These substances are moved with an appropriate carrier system. Both of the most extensively studied active transport systems use the high-energy molecule adenosine triphosphate (ATP).

In some cases, it appears that specific globular proteins within the cell membrane temporarily change shape under certain conditions. The modification may temporarily open a pore wide enough for the transported molecule to pass through. When active transport is involved, energy from ATP may help change the conformation of the **carrier molecule** so it will pick up the molecule that is to be moved, or ATP may help detach the transported molecule from its carrier once it has been moved to the other side of the membrane (see Figure 2.4).

Movement of Large Particles across Membranes

The plasma membrane has the capacity to engulf small particles on either side of the cell and then fuse with these particles; or sometimes the portion of the plasma membrane that surrounded the small particles simply separates, pinching itself off and drifting away. When liquids or macromolecules are engulfed by the membrane and are then moved to the other side, the process is known as **pinocytosis**. The same process, but on a larger scale, that takes on larger particles is **phagocytosis**. These engulfed particles pinch off with part of the membrane and move in or out of the cell. The process that moves particles into the cell is **endocytosis**, and the process that moves particles out of the cell is **exocytosis**.

PLANT CELL WALL

All cells have a plasma membrane, but some also produce a nonliving outer **cell wall**. The cell wall is a product of the cytoplasm that supports and provides shape for the cell. Most bacterial cells have a rigid or semirigid cell wall outside the plasma membrane that is usually composed of a polymer of glucose derivatives attached to amino acids. Sometimes, certain bacteria have an additional layer outside this known as lipopolysaccharide (a polymer composed of lipid and sugar monomers).

The complex polysaccharide that composes most plant cell walls is **cellulose** (and related compounds). The cellulose wall allows a wide range of compounds to pass through easily. It is the cell membrane that determines which materials may enter or leave the cell.

In complex, multicellular plants, the **primary cell wall** is the outer layer that is laid down first by the young cells. And in some of these plants, where two cell walls come in contact, another layer is laid down between the primary cells walls. This is known as the **central** or **middle lamella**, which is one continuous layer between the cells, and it is usually composed of **pectin**, a complex polysaccharide, generally in the form of calcium pectate. It is the pectin that binds the cells together.

Plant cells that become harder, or more woody, usually do so by adding layers to the cell wall, known as the **secondary wall**. This wall is also deposited by the cytoplasm and is inside the primary wall. Rather than being laid down in a loose network, as in the primary cell wall, the long threadlike structures that make up much of the cell wall, known as **cellulose fibrils**, are

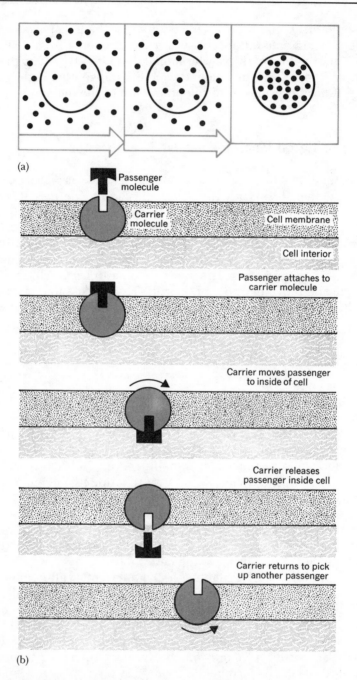

(a)

(b)

Figure 2.4 *Active Transport: a) A cell concentrates certain kinds of molecules inside, or may move them outside the membrane, utilizing a process that expends energy; b) the process works with carrier molecules as illustrated.*

laid down parallel to one another in the secondary cell wall in a series of layers, or lamellae. Each layer has fibrils oriented at angles of about 60 degrees to the fibrils of the next lamellae. The parallel fibrils, the number of layers, their orientation, the total thickness, as well as added substances such as **lignin**, help strengthen the cell wall.

PLASMODESMATA AND GAP JUNCTIONS

Plasma membrane connections, known as **desmosomes**, are thin cytoplasmic connections that often run from one plant cell to the next through pores in the cell walls. These pores, called **plasmodesmata**, facilitate intercellular exchange of materials such as sugars and amino acids (see Figure 2.5).

Gap junctions, sometimes called **nexus junctions**, are areas of low electrical resistance between animal cells across which electrical impulses, small ions, and molecules may pass. They seem to function similarly to the plasmodesmata that connect plant cells.

When animal cells are in water with a low relative osmotic pressure, water will move from the area of low osmotic pressure into the cells with higher osmotic pressure, and they will ultimately burst due to the cells' distention, or **turgor**. Plant cells can usually withstand the swelling without bursting because their sturdy cell walls counterbalance the low external pressure.

CELL'S INTERNAL STRUCTURES

Nucleus

Relatively large, distinct, membrane-enclosed structures of several kinds are found in most kinds of cells, though not all. The notable exceptions are bacteria and cyanobacteria (often referred to interchangeably as blue-green bacteria or blue-green algae). The largest, most well-defined subcellular structure, or **organelle**, is the nucleus, unless even larger **vacuoles** are present. Vacuoles are large, fluid-filled spaces found in many cells, particularly plant cells. The nucleus contains the long thin structures composed of **deoxyribonucleic acid (DNA)** and protein—the **chromosomes**. These structures contain **genes**, which are the individual units of information that inform the cell what to do and how and when to do it. All the instructions concerning the life processes of the cell emanate from these chromosomes (see Figure 2.6).

Nucleic acids are a major group of organic compounds that are composed of subunits called **nucleotides**, which contain carbon, hydrogen, oxygen, phosphorus, and nitrogen. It is the sequence of nucleotides that gives the specific nucleic acid its distinctive properties. Deoxyribonucleic acid and **ribonucleic acid (RNA)** are two important nucleic acids found in cells. They are fundamental in the storage and transmission of genetic information that controls the cell's functions and interactions.

(a)

(b)

Figure 2.5 *Cells have specialized junctions that facilitate the passage of materials across plasma membranes. Shown in (a) are the surfaces of sweat gland cells, and (b) desmosomes, which form below these surfaces in plasma membranes of adjacent cells.*

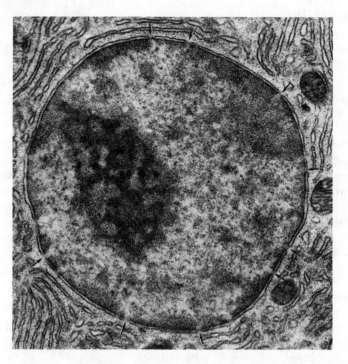

Figure 2.6 *An electron micrograph of a nucleus. The arrows point to pores in the nuclear envelope's double membrane. The large dark area in the left portion of the nucleus is its nucleolus. Most of the other dark material inside the nucleus is chromatin.*

A dark area within the nucleus of most cells, seen when the cells are not dividing, is known as the **nucleolus** (see Figure 2.6). This area contains a high concentration of **ribonucleic acid (RNA)** and protein. Sometimes a nucleus contains several nucleoli, which are formed by a specific region on one chromosome called the **nucleolar organizer**.

All the nuclear structures are embedded in a viscous colloidal material, the **nucleoplasm**. The nucleus is bounded by a membrane, the **nuclear membrane**, which consists of two complete membranes similar in structure to the cell membrane. The nuclear membrane is very selective, restricting certain substances that readily pass through the cell membrane into the cytoplasm but are unable to pass through the nuclear membrane.

The nuclear membrane is perforated by many large pores that facilitate the passage of substances in and out of the nucleus. Attached to the outside of the nuclear membrane are many thin membranes of **endoplasmic reticulum** that appear to be important in assisting the two-way flow of materials to and from the nucleus.

Endoplasmic Reticulum

The **endoplasmic reticulum (ER)** is a series of double-layered membranes found throughout the cytoplasm that provide surface area for enzymes to catalyze chemical reactions, particularly those leading to protein synthesis. The ER forms a multibranched channel between the outside of the cell and the nucleus, connecting the cell membrane and the nuclear membrane, thus expediting the transport of cellular materials.

Endoplasmic reticulum may be present with or without the outer surface lined with small particles called **ribosomes**, which are the actual sites of protein synthesis. Endoplasmic reticulum without ribosomes is known as **smooth endoplasmic reticulum**, and this is where some steroid hormones and other lipids are synthesized. The endoplasmic reticulum that is lined with ribosomes is referred to as **rough endoplasmic reticulum**. These terms, smooth and rough endoplasmic reticulum, relate to the structures' appearance in electron micrographs (see Figure 2.7).

Ribosomes

Ribosomes are tiny granules that can exist either in the cytoplasm as free-floating organelles or along the outer surface of the endoplasmic reticulum. Cells contain three main types of ribonucleic acid (RNA), each synthesized from DNA. The RNA found in ribosomes that functions in protein synthesis is known as **rRNA**, which stands for **ribosomal RNA**.

Ribosomes contain high concentrations of rRNA and protein. In addition to existing singly, and attached to endoplasmic reticulum, some ribosomes occur in clumps or clusters, which are collectively known as **polyribosomes** or **polysomes**.

Golgi Apparatus

Another type of cellular organelle is the **Golgi apparatus**. These typically look like a stack of flattened, smooth, oval membranes. It is thought that certain chemicals that are synthesized in the ribosomes move along the endoplasmic reticulum to the Golgi apparatus, where they are concentrated and stored until released by secretory vesicles produced by the outer portion of the Golgi apparatus.

In addition to functioning as an area where substances such as proteins pass through and are stored before being secreted, the Golgi apparatus is also involved in processing some proteins and carbohydrates, synthesizing them into more complex molecules by combining them into **glycoproteins** and **mucopolysaccharides**, some of which are secreted in the form of mucus (see Figure 2.8).

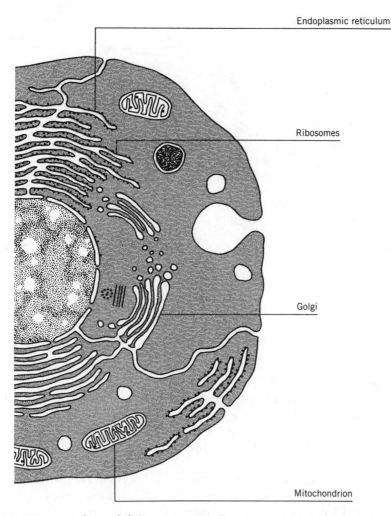

Endoplasmic reticulum

Ribosomes

Golgi

Mitochondrion

Figure 2.7 *Here are four of the major organelles located in the animal cell as they would appear under an electron microscope. Note that both smooth and rough endoplasmic reticulum are shown.*

Lysosome

A **lysosome** is a membrane-enclosed storage vesicle. The membrane is composed of a single layer, unlike the double-layered membrane of the nucleus. Lysosomes appear to arise from Golgi bodies. They contain strong digestive (hydrolytic) enzymes that act as the cell's digestive system, making it possible for the cell to process materials within its cytoplasm such as the particles taken in by pinocytosis or phagocytosis (defined earlier in this chapter under the heading "Movement of Large Particles across Cell Membranes").

Lysosomes also break down old parts of the cell into the organic molecules from which they were constructed. These molecules are then reused by the

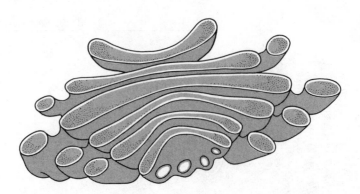

Figure 2.8 *This cross section diagram illustrates the Golgi apparatus, an organelle that provides capsules in which hormones and digestive enzymes are stored.*

cell. The importance of lysosomes has been established in studies concerning the metamorphosis of frog larvae. When tadpole tails shrink in size, their constituents are being broken down by lysosomes, and these organic molecules are then reabsorbed to fuel the growth of new structures such as the frog's budding arms and legs.

Mitochondria

Often referred to as the powerhouse of the cell, **mitochondria** supply the cell with chemical energy in the form of the high-energy molecule adenosine triphosphate (ATP). Slightly larger than the lysosomes, these double-membraned organelles (see Figure 2.9) are often the largest of the organelles in the cytoplasm, although **plastids**, **vacuoles** (described later), and the nucleus are also quite large in many cases.

Because they supply cells with energy, mitochondria are most abundant in cells with high energy requirements, as in the growing root tip of a plant or in an animal's muscle cell. Each liver cell contains as many as 2,500 mitochondria.

The outer membrane separates the mitochondria from the cytoplasm, and the inner membrane has many inwardly directed folds, or **cristae**, that extend into a semifluid **matrix**. It is along the cristae and in the matrix where the energy-transfer enzyme systems that manufacture ATP are found. ATP is used in cell division, muscle cell contraction, and as an energy source for many other cellular activities throughout living organisms.

When more sugar is available in a cell than it requires, **fat droplets** found next to many mitochondria will produce fat. Then when an energy deficit arises, the mitochondria will break down the fat droplets to yield energy. Excess fat droplets are transferred to specialized fat storage cells.

Unlike most other organelles, mitochondria contain DNA, RNA, and ribosomes, and they replicate themselves. It is now thought that mitochondria did not arise as a part of the cell but were originally independently living cells

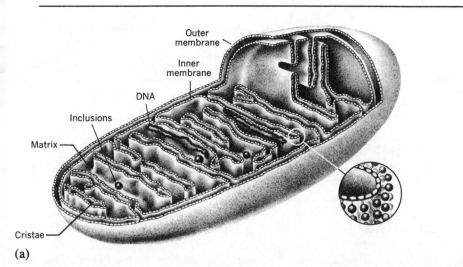

(a)

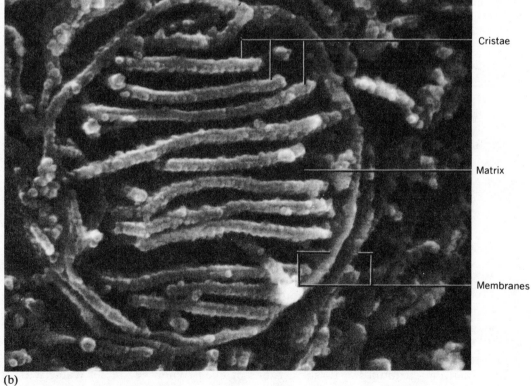

(b)

Figure 2.9 *a) An artist's rendering of a mitochondrion; and b) a labeled photomicrograph of a mitochondrion in cross section.*

that were taken into larger cells and were able to survive in a jointly beneficial relationship (or **symbiosis**).

Mitochondria are often found in close proximity with a smaller organelle, the **peroxisome**, which may produce substrates for mitochondrial activity.

Plastids

Like mitochondria, **plastids** contain DNA, RNA, and ribosomes and are able to replicate themselves. Plastids are rather large cytoplasmic organelles that can be seen with a light microscope. They occur in most photosynthetic plant cells and also in photosynthetic protists (single-celled organisms such as some algae). Plastids are composed of two membrane layers. The **outer membrane** lacks folds and encloses the organelle, regulating the movement of materials between the cytoplasm and the organelle's interior. The **inner membrane** folds throughout the internal proteinaceous matrix known as the **stroma**. These internal membranes occur in sheets called **lamellae**, which often form stacks or **grana**; these stacks are connected by single lamellae.

Different types of plastids have been recognized, and each appears to develop from a common structure or sometimes from one another. Those that are white and colorless and that store oils and protein granules as well as energy-rich starch formed from simple sugars are the **leucoplasts**. Plastids filled with starch are called **amyloplasts**. Plastids containing yellow, orange, or red pigments are called **chromoplasts**. These are found in flowers and fruits. **Chloroplasts** are the chromoplasts that–in addition to yellow, orange, or red pigments–also contain green pigment, or **chlorophyll**, which uses energy from sunlight to manufacture organic molecules such as sugar from inorganic molecules. In addition to chlorophyll, some of the other pigments, including **carotenoids** and **xanthophylls**, remain in temperate zone leaves in the autumn after the green pigments have broken down. It is these pigments that account for the autumn colors (see Figure 2.10).

Vacuoles

Vacuoles are membrane-bound, fluid-filled spaces in both plant and animal cells, although they are most prominent in some plant cells (see, for instance, Figures 2.11 and 2.12). Much of the total volume of most plant cells is composed of a single vacuole surrounded by its membrane, the **tonoplast**. The liquid inside the plant vacuole is called **cell sap**, which is mostly water with some dissolved substances. Cell sap is usually hyperosmotic relative to the external medium surrounding it, so in times of need, when available, water will move in through the vacuole's membrane via osmosis, keeping the cell turgid. Vacuoles often contain a high concentration of soluble organic compounds such as sugars, amino acids, some proteins, and several pigments. Many plant species store toxic chemicals in the vacuoles that do not harm the cell. If an

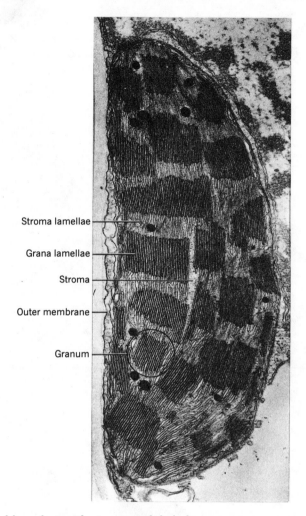

Stroma lamellae

Grana lamellae

Stroma

Outer membrane

Granum

Figure 2.10 A chloroplast with structures labeled.

animal eats the plant, however, the contents of the ruptured vacuoles can have an adverse effect on the animal.

Different kinds of vacuoles have different functions. Some single-celled organisms have **contractile vacuoles**, which are important in expelling excess water and waste from the cell. Many cells have vacuoles or vesicles that are formed when the cell membrane takes in small particles by phagocytosis or pinocytosis, which may then be digested by enzymes from the lysosomes.

Microfilaments

Within a cell's cytoplasm are many long protein fibers arranged into **microfilaments** that are only about six nanometers in diameter. Microfilaments func-

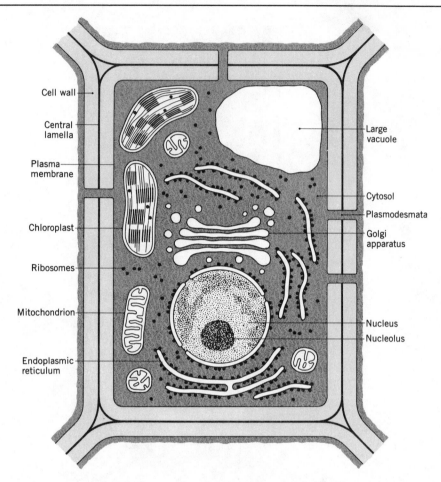

Cell wall

Central
lamella

Plasma
membrane

Chloroplast

Ribosomes

Mitochondrion

Endoplasmic
reticulum

Large
vacuole

Cytosol

Plasmodesmata

Golgi
apparatus

Nucleus

Nucleolus

Figure 2.11 *A generalized plant cell with its structures labeled.*

tion in the maintenance of cell structure and movement. Depending on the local conditions in the cell, microfilaments can exist either as individual subunits or together in strings of subunits. Microfilaments are believed to be important components of the contractile filaments that exist in some cells. They may be associated with **cytoplasmic streaming**, a constant movement of cytoplasm in the cell that is evident in many plant cells under the light microscope. Microfilaments may also be involved in cell movement, and when in a more complex form, they may be associated with the contraction of some muscle cells.

Microtubules

Also in the cytoplasm are **microtubules** that differ from microfilaments in many respects. The outside diameter of each microtubule is considerably larger than that of a microfilament, ranging from 20 to 25 nanometers. Furthermore,

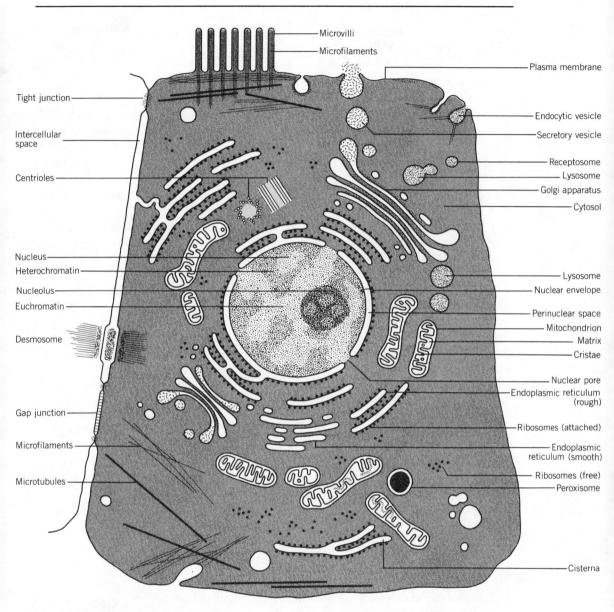

Figure 2.12 *Composite animal cell.*

the microtubules are hollow and vary in length from less than 200 nanometers to 25,000 nanometers. Microtubules also differ from microfilaments in terms of their chemical make up.

Sometimes, microtubules act as tracts along which the cell's organelles move. Microtubules orient themselves toward each pole of a cell, and chromosomes move along them during cell division. A distinctive internal microtubular arrangement also determines the arrangement of a cell's **cilia, flagella,** and **sperm tails**, which are discussed below.

Cilia and Flagella

Cilia and **flagella** (singular: cilium and flagellum) are long, thin organelles that project from the surface of some cells and have the capacity to beat back and forth or in a corkscrew fashion, moving the cell through fluids, or moving fluids past a stationary cell. They appear on some protists, on certain lower plants, and on some animals. Many eukaryotic organisms have flagellated sperm cells. Eukaryotic and prokaryotic bacterial flagella differ structurally and mechanically.

The difference between cilia and flagella is mostly a matter of length (though not diameter), and sometimes they are differentiated according to number; many short ones together are called cilia, while an isolated long one is usually a flagellum.

The function of these organelles is primarily locomotory, to propel or pull a cell through liquids. Some protists also use cilia to move particles into their gullet—a part of the cell membrane that is similar to a mouth. Multicellular organisms commonly have ciliated cells lining internal passageways. Their constant motion helps move materials through tubes and ducts in the preferred direction.

In all types of cells except bacteria and blue-green algae, cilia and flagella are cytoplasmic outgrowths from the surface of the cell, surrounded by extensions of the plasma membrane. At the base of the long cilium or flagellum is the **basal body**, which supports and gives rise to the rest of the long stalk. It consists of nine triplets of microtubules arranged in a circle. The basal bodies may be where the protein subunits assemble to form the microtubules that extend through the rest of the long structures, which are themselves composed of a circle of nine pairs of microtubules with another pair at their center. This 9 + 2 arrangement is characteristic of all cellular cilia and flagella except those found among bacteria and blue-green algae.

Centrioles

Basal bodies and **centrioles** are structurally identical, although the basal bodies are found at the bases of cilia and flagella of all cells, while centrioles are found only in animal cells, generally located near the nucleus. Centrioles occur in pairs. Like basal bodies, they are composed of nine sets of triplets, all arranged lengthwise, creating a cylindrical structure.

Not all cells contain centrioles, but most animal cells have them, and they are always composed of the same typical pattern: a cylinder formed by nine sets of triplets that, when cut in a transverse section (cross section), looks like a pinwheel (see Figure 2.13).

Besides playing a role in the formation of microtubules, centrioles become active during cell reproduction. During interphase, both pairs of centrioles are located just outside the nucleus, arranged at right angles to each other. As

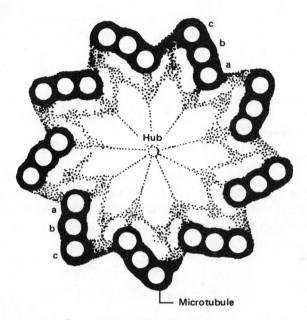

Figure 2.13 *Cross section of a centriole. This drawing shows the nine triplets of micro-tubules, which turn in unison, creating the corkscrew or whiplike motion of cilia and flagella.*

mitosis (or meiosis) begins, they move apart and organize the mitotic apparatus known as the **spindle**, which is composed of microtubules extending from pole to pole through the cell's interior. It is along these microtubules that the chromosomes align at metaphase, then moving apart during the remaining stages of cell division. Toward the end of cell division (during telophase), the centrioles replicate, so each daughter cell in turn has two pairs of centrioles located just outside the nucleus. For a more comprehensive discussion of cell division, see Chapter 3.

Eukaryotic and Prokaryotic Cells

The differences between cells of bacteria and blue-green algae and those of most other organisms are customarily cited to distinguish between two distinct cellular types: eukaryotic cells and prokaryotic cells.

Eukaryotic Cells

Most of the preceding part of this chapter concerns eukaryotic cells. Eukary-otes are those plants and animals with eukaryotic cells. These include all the protists, fungi, plants, and animals. Their cells have a nucleus enclosed within a nuclear membrane, inside of which are located the **chromosomes**—the long, filamentous structures along which occur the units of inheritance, called **genes**. These chromosomes are composed of nucleic acids and proteins.

Eukaryotic cells also contain mitochondria, Golgi bodies, endoplasmic reticulum, and most of the other organelles so far described. When present in these cells, chlorophyll is located in chloroplasts. And if these cells have cilia and/or flagella, they always possess the previously described 9 + 2 construction.

All the contents of eukaryotic cells are enclosed within a plasma membrane, and in some cases, as with many plant cells, the plasma membrane is also enclosed within a cell wall.

Prokaryotic Cells

Prokaryotic cells are thought to be more primitive than eukaryotic cells. Bacteria and blue-green algae are **prokaryotes**, which is to say their cells are prokaryotic. These cells do not have a nuclear membrane; instead, they have a nuclear area known as the **nucleoid** that in most cases contains a single circular chromosome. In prokaryotic cells, the chromosomes are composed entirely of nucleic acids (without protein). Their ribosomes are also structurally distinct from those in eukaryotic cells. When flagella-like structures are present, they don't have the 9 + 2 fibrillar structure. And prokaryotic cells contain no mitochondria, no Golgi apparatus, no lysosomes, and no vacuoles.

In addition, prokaryotic cells have cell walls containing muramic acid, a characteristic not found among eukaryotic cells. In many prokaryotic cells appears an infolding of the plasma membrane known as a **mesosome**, which may be involved in the production of new cell wall material following cell division, and which may be associated with the replication of DNA during cell division. The mesosome may also help break down food molecules to supply the cell with energy.

Most photosynthetic prokaryotes contain photosynthetic membranes and chlorophyll, but these materials aren't enclosed in chloroplasts.

Key Terms

active transport
amyloplasts
basal body
carotenoids
carrier molecules
cell membrane
cell sap
cell theory
cell walls
cells
cellulose
cellulose fibrils
centriole
chemical homeostasis
chlorophyll
chloroplasts

cilia
colloid
concentration gradient
contact X-ray microscopy
contractile vacuoles
cristae
cytologists
cytoplasm
cytoplasmic streaming
deoxyribonucleic acid
diffusion
dissecting microscopes
electron microscopes
endocytosis
endoplasmic reticulum
equilibrium

eukaryotic cells
exocytosis
facilitated diffusion
fat droplets
flagella
gap junctions
gel
genes
glycoprotein
Golgi apparatus
grana
Hooke, Robert
hyperosmotic
hypertonic
hypoosmotic
hypotonic
inner membrane
isosmotic
isotonic
lamellae
Leeuwenhoek, Antonie van
leucoplasts
light microscope
lignin
lysosome
magnification
matrix
mesosome
microfilaments
microscope
microtubules
middle lamella
mitochondria
mucopolysaccharides
multicellular organism
nexus junctions
nuclear membrane
nucleic acids
nucleoid
nucleolar organizer
nucleolus
nucleoplasm
nucleotides

nucleus
organelle
osmosis
osmotic pressure
outer membrane
passive transport
pectin
peroxisome
phagocytosis
pinocytosis
plasma membrane
plasmodesmata
plastids
polyribosomes
polysomes
primary cell wall
prokaryotes
prokaryotic cells
resolution
ribonucleic acid
ribosomal RNA
ribosome
rough endoplasmic reticulum
scanning tunneling microscope
Schleiden, Matthias
Schwann, Theodor
secondary wall
semipermeable membrane
smooth endoplasmic reticulum
sol
solution
sperm tails
spindle
stereomicroscopes
stroma
symbiosis
tonoplast
transmission electron microscope
turgor
unicellular organism
vacuole
xanthophylls

Chapter 2 Self-Test

QUESTIONS TO THINK ABOUT

1. Explain the events that made it possible to observe and understand cell structure and function.
2. When cells were first described, what was observed during this initial discovery?
3. Describe what is now known as cell theory.
4. Why are most cells small?
5. Define *cytoplasm* and *nucleoplasm*.
6. Compare and contrast the function of a solution and a colloid in cell structure.
7. What is a cell membrane?
8. How do materials move through a cell membrane?
9. What are the possible dynamics of cell contents when the osmotic pressure within the cell is not equal to that outside the cell?
10. Compare and contrast passive transport and active transport.
11. How do materials move between cells?
12. Describe the nucleus: what is it; where is it; what is in it; and what is its function?
13. Give a brief description of each of the following structures:
    a. endoplasmic reticulum
    b. ribosome
    c. Golgi apparatus
    d. lysosome
    e. mitochondrion
    f. fat droplet
    g. plastid
    h. vacuole
    i. microfilament
    j. microtubule
    k. cilium
    l. flagellum
    m. centriole
14. Compare and contrast eukaryotic and prokaryotic cells.

MULTIPLE-CHOICE QUESTIONS

Introduction, Cells, Cell Theory

1. During the seventeenth century, the first light microscopes were constructed by _____ .

   a. Hooke
   b. DuPont
   c. Scopes
   d. Pierson
   e. Leeuwenhoek

2. The study of cells is called _____ .

   a. herpetology
   b. ornithology
   c. simology
   d. cytology
   e. tautology

3. Illuminating specimens under the light microscope with _____ _____ light was found to give considerably more resolution than when using other colors.

   a. red
   b. orange
   c. blue
   d. green
   e. yellow

4. With a transmission electron microscope, a beam of electrons with considerably _____ than visible light is passed through a thinly sliced specimen.

   a. longer wavelengths
   b. shorter wavelengths
   c. brighter light
   d. stronger X-rays
   e. stronger cosmic rays

5. Of the following, the most recent advance in microscope technology that enables scientists to observe the most highly detailed views of individual living cells is called _____ .

   a. transmission electron microscopy
   b. light microscopy
   c. scanning electron microscopy
   d. contact X-ray microscopy
   e. gamma ray microscopy

6. In 1665 _____ described having seen what he called cells in a piece of cork.

    a. Spallanzini
    b. Malthus
    c. Hooke
    d. Leeuwenhoek
    e. Marvin

7. All following statements concerning cell theory except one are correct. Circle the letter preceding the incorrect answer.

    a. Cells are the fundamental units of life.
    b. All cells have cell walls.
    c. Cells are the smallest entities that can be called living.
    d. All organisms are made up of one or more cells.
    e. Cells arise only from the division of other cells.

Cell Structure and Cell Size, Cytoplasm or Protoplasm, Plasma Membrane

8. A cell becomes too large to maintain a stable internal environment when

    a. its surface area becomes too small
    b. its volume becomes too small
    c. its surface area gets too big
    d. its volume gets too big
    e. its surface area to volume ratio becomes too great

9. It has been stated that the larger the cell, the _____ to maintain control over the entire internal environment.

    a. easier it is for the nuclear material
    b. harder it is for the nuclear material
    c. easier it is for the chloroplast
    d. harder it is for the chloroplast
    e. easier it is for the mitochondria

10. The interior of a living cell is made up of a _____ .

    a. complex solution
    b. a heterogeneous colloid
    c. a solid throughout
    d. a and b
    e. b and c

11. A colloid may become a gel, depending on the _____ .

    a. temperature
    b. pH
    c. salt concentration
    d. pressure
    e. all of the above

Movement through the Cell Membrane, Diffusion, Osmosis,
Facilitated Diffusion

12. The plasma membrane provides the following to the cell:

   a. protection
   b. shape
   c. strength
   d. a structure that helps regulate the flow of materials going in and out of the cell
   e. all of the above

13. When energy is used to move molecules from one location to another, it is termed _____ .

   a. passive transport
   b. colloidal collision
   c. gel state
   d. passive homeostasis
   e. active transport

14. The net movement of particles from regions of higher concentration to regions of lower concentration of that substance is called _____ _____ .

   a. homeostasis
   b. active ambiance
   c. homeostatic collision
   d. diffusion
   e. gradual constance

15. The particles moving away from the area of greatest concentration of that substance will produce a _____ .

   a. gel state
   b. colloidal collision
   c. diffusionary divergence
   d. gradual homeostasis
   e. concentration gradient

16. A partially or differentially permeable membrane is called a _____ _____ .

   a. diffusionary membrane
   b. semihomeostatic membrane
   c. active membrane
   d. facilitated membrane
   e. semipermeable membrane

17. The pressure exerted by dissolved particles in a solution or a colloid that pulls water across a semipermeable membrane is known as _____ .

    a. isosmosis
    b. hyperosmosis
    c. osmosis
    d. hypoosmosis
    e. osmotic pressure

18. When there is a semipermeable membrane separating two solutions of pure water, the two sides are at _____ .

    a. hyperequilibrium
    b. hypoequilibrium
    c. osmotic diffusion
    d. facilitated osmoequilibrium
    e. equilibrium

19. When the osmotic pressure is the same on both sides of a semipermeable membrane, each solution is said to be _____ .

    a. isosmotic
    b. hyperosmotic
    c. hypoosmotic
    d. osmotic
    e. none of the above

20. Plasma membranes contain _____ that help move certain substances in and out of the cell.

    a. organelles
    b. nuclei
    c. nucleoli
    d. prions
    e. carrier molecules

Active Transport, Movement of Large Particles Across Membranes, Plant Cell Wall

21. To move substances such as glucose across a cell membrane against a concentration gradient _____ .

    a. no energy needs to be expended
    b. is called diffusion
    c. is called osmosis
    d. energy has to be expended
    e. none of the above

22. The most extensively studied active transport systems use the high-energy molecule ———————— to maintain the systems.

    a. ATP
    b. $NADPH_2$
    c. NAD
    d. $NADH_2$
    e. none of the above

23. The connected network of similar membranes that connect structures within the cell are called (the) ———————— .

    a. carrier molecules
    b. endoplasmic reticulum
    c. globular proteins
    d. adenosine triphosphate
    e. gamma globulin

24. The capacity of the plasma membrane to fuse with or engulf particulate matter and then pinch off and move out of the cell is known as

    ———————— .

    a. endocytosis
    b. exocytosis
    c. endothelial phagocytosis
    d. exocrine pinocytosis
    e. none of the above

25. When liquids or macromolecules are engulfed and moved into the cell, this kind of endocytosis is known as ———————— .

    a. exocytosis
    b. endothelial phagocytosis
    c. pinocytosis
    d. phagocytosis
    e. none of the above

26. Both plant and animal cells possess a plasma membrane, but only plant cells have a ———————— .

    a. mesothelium
    b. endothelium
    c. cell wall
    d. nucleus
    e. nucleolus

27. The cell wall is a product of the cellular protoplasm and is primarily composed of the complex polysaccharide _____ and other related compounds.

    a. starch
    b. glucose
    c. globular protein
    d. cellulose
    e. nylon

28. Where two cell walls come in contact, another layer is laid down between the two cells, which is called the middle lamella. This continuous layer between the cells is usually composed of _____ .

    a. starch
    b. pectin
    c. lignin
    d. globular protein
    e. cellulose

29. The complex polysaccharide, usually in the form of calcium pectate, that binds plant cells together, is known as _____ .

    a. cellulose
    b. lignin
    c. beta kerotene
    d. anthracnose
    e. pectin

30. The long threadlike structures that make up much of the cell wall, and that are laid down in a loose network, are known as _____ .

    a. myofibrils
    b. myoglobin
    c. fibrinogen
    d. cellulose fibrils
    e. pectin

Plasmodesmata, Gap Junctions, Nucleus, and Endoplasmic Reticulum

31. The thin protoplasmic connections that often run from one plant cell to the next through pores in the cell walls are known as _____ .

    a. plasmids
    b. plasmodesmids
    c. gap junctions
    d. endoplasmic reticulum
    e. plasmodesmata

32. When animal cells are in water with a low osmotic pressure relative to their interior, the water will move from the area of low osmotic pressure into the cells with the higher osmotic pressure, and ultimately the cells will burst due to their distension, or _____ .

    a. plasma
    b. desmata
    c. turgor
    d. nexus
    e. none of the above

33. Well-defined subcellular structures that are enveloped in membranes are called _____ .

    a. bacteria
    b. blue-green algae
    c. prions
    d. nucleotides
    e. organelles

34. The large fluid-filled spaces enveloped in membranes that are found in many cells, particularly in plant cells, are _____ .

    a. chromosomes
    b. vacuoles
    c. nuclei
    d. chromatids
    e. nucleoli

35. The large organelle containing chromosomes is the _____ .

    a. nucleus
    b. vacuole
    c. chromatid
    d. lysosome
    e. endoplasmic reticulum

36. It is the sequence of _____ that gives the specific nucleic acid its distinctive properties.

    a. genes
    b. nuclei
    c. chromosomes
    d. nucleotides
    e. nucleoli

37. When the cell is not dividing, there is a conspicuous dark area within the nucleus, the _____ , that contains a high concentration of ribonucleic acid and protein.

    a. endoplasmic reticulum
    b. Golgi body
    c. vacuole
    d. lysosome
    e. nucleolus

38. All the nuclear structures are embedded in a viscous colloidal material, the _____ .

    a. RNA
    b. DNA
    c. cytoplasm
    d. protoplasm
    e. nucleoplasm

39. The nucleus is bounded by the _____ , which consists of two complete membranes similar in structure to the plasma membrane.

    a. nuclear membrane
    b. nuclear wall
    c. nucleolar organizer
    d. nucleoplasm
    e. cell wall

40. Attached to the outside of the nuclear membrane are many other membranes of _____ that appear to be important in assisting the two-way flow of materials to and from the nucleus.

    a. cell wall
    b. endoplasmic reticulum
    c. endothelium
    d. nuclear organizer
    e. cell membrane

41. The _____ is a series of double-layered membranes found throughout the cytoplasm that greatly increase the surface area within the cytoplasm, expanding the area available for enzymes involved in chemical reactions, particularly those leading to protein synthesis.

    a. endoplasmic reticulum
    b. primary cell wall
    c. cell wall
    d. cell membrane
    e. plasma membrane

42. Endoplasmic reticulum may occur with and without being lined with small particles, the _____ , which are the actual sites of protein synthesis.

    a. ribosomees
    b. lignin
    c. pectin
    d. endothelium
    e. nuclei

Ribosomes, Golgi Apparatus, Lysosomes, Mitochondria, and Plastids

43. Besides being found in cells singly as well as on endoplasmic reticulum, some ribosomes occur in clumps or clusters collectively known as _____ .

    a. polysomes
    b. polyribosomes
    c. polychromosomes
    d. a and b
    e. a, b, and c

44. Certain _____ synthesized in the ribosomes move along the endoplasmic reticulum to the _____ , where they are concentrated.

    a. plastids, lysosomes
    b. peroxisomes, autosomes
    c. glycoproteins, mesosomes
    d. polysaccharides, mesosomes
    e. chemicals, Golgi apparatus

45. Chemicals that are concentrated and stored in the Golgi apparatus are later released from the cell by secretory vesicles that are produced by the outer portion of the Golgi apparatus when it moves to the _____ .

    a. cell surface
    b. lysosome
    c. mesosome
    d. autosome
    e. ribosome

46. Some simple sugars may be modified in the Golgi apparatus, synthesizing more complex polysaccharides that can then be attached to proteins forming _____ and _____ that can be used for mucus secretion.

    a. starch, celluloproteins
    b. glycogen, mucoproteins
    c. photoglycogen, mucoglycogen
    d. glycopolysaccharides, glucopolysaccharides
    e. glycoproteins, mucopolysaccharides

47. _____ contain strong digestive enzymes that act as the cell's digestive system, making it possible for the cell to process materials within its cytoplasm such as the particles taken in by pinocytosis or phagocytosis.

    a. peroxisomes
    b. lysosomes
    c. mesosomes
    d. autosomes
    e. ribosomes

48. Often referred to as the powerhouse of the cell, _____ supply the cell with the chemical energy in the form of the high energy molecule adenosine triphosphate.

    a. mitochondria
    b. vacuoles
    c. endoplasmic reticulum
    d. Golgi apparatus
    e. mesosomes

49. The inner membranes of the mitochondria have many inwardly directed folds, the _____ , which extend into a semifluid matrix.

    a. cristae
    b. grana
    c. thylakoids
    d. lamellae
    e. stroma

50. In plastids, stacks of lamellae are the _____ .

    a. stroma
    b. amyloplasts
    c. chloroplasts
    d. leucoplasts
    e. grana

51. The many double membranes found in plastids are called _____ .

    a. stroma
    b. grana
    c. lamellae
    d. amyloplasts
    e. leucoplasts

52. The proteinaceous matrix inside plastids is known as the _____ .

    a. stroma
    b. grana
    c. amyloplasts
    d. leucoplasts
    e. thylakoids

53. Plastids filled with starch are called _____ .

    a. leucoplasts
    b. amyloplasts
    c. chromoplasts
    d. leucocytes
    e. granulocytes

54. Most plastids that contain yellow, orange, or red pigments are called

    _____ .

    a. leucoplasts
    b. amyloplasts
    c. chromoplasts
    d. leucocytes
    e. granulocytes

55. The chromoplasts that contain green pigment are known as _____

    _____ .

    a. leucoplasts
    b. amyloplasts
    c. chromoplasts
    d. chloroplasts
    e. carotenoplasts

56. In addition to chlorophyll, chloroplasts also contain other pigments, particularly the yellow and orange _____ .

    a. amylophyll
    b. leucophyll
    c. rohophyll
    d. chromophyll
    e. carotenoids

Vacuoles, Microfilaments, Microtubules, Cilia and Flagella, Centrioles, and Eukaryotic and Prokaryotic Cells

57. Much of the total volume of most plant cells is comprised of a single

    _____ .

    a. vacuole
    b. tonoplast
    c. ribosome
    d. autosome
    e. macrotubule

58. Vacuoles are surrounded by a membrane known as the _____ .

    a. microtubule
    b. macrotubule
    c. microfilament
    d. tonoplast
    e. amyloplast

59. The liquid inside the vacuole is called _____ .
    a. cell sap
    b. tonoplasm
    c. vacuoplasm
    d. a and b
    e. a and c

60. Membrane-bound fluid-filled spaces in plant and animal cells, which
    in some plant cells comprise much of the total cell volume, are
    _____ .
    a. tonoplasts
    b. vacuoles
    c. ribosomes
    d. mesosomes
    e. autosomes

61. There are different kinds of vacuoles with various functions. Some single-
    celled organisms have _____ that are important in expelling
    excess water and waste from the cell.
    a. tonoplasts
    b. ribosomes
    c. autosomes
    d. contractile vacuoles
    e. mesosomes

62. Within a cell's cytoplasm are many long, thin protein fibers that are only
    about six nanometers in diameter. Depending on the local conditions in the
    cell, these _____ can exist either as subunits or as filaments.
    a. cilia
    b. microfilaments
    c. flagella
    d. basal bodies
    e. microtubules

63. The constant movement of cytoplasm in the cell is called _____
    _____ .
    a. cytoplasmic streaming
    b. ciliar movement
    c. flagellar movement
    d. microtubular movement
    e. microfilamentary streaming

64. Cilia and flagella are cytoplasmic outgrowths from the surface of the cell, surrounded by extensions of the cell membrane. The base of the long cilium or flagellum is the ─────────── .

    a. centriole
    b. microtubule
    c. microfilament
    d. basal body
    e. blue-green algae

65. During interphase in animal cells, but not in most plant cells, in a special region of cytoplasm just outside the nucleus are two cylindrical bodies oriented at right angles to each other that move apart and organize the mitotic apparatus of the dividing cell. These organelles are the ─────────── .

    a. basal bodies
    b. microtubules
    c. microfilaments
    d. centrioles
    e. cilia

ANSWERS

1. e	18. e	35. a	52. a
2. d	19. a	36. d	53. b
3. c	20. e	37. e	54. c
4. b	21. d	38. e	55. d
5. d.	22. a	39. a	56. e
6. c	23. b	40. b	57. a
7. b	24. b	41. a	58. d
8. e	25. c	42. a	59. a
9. b	26. c	43. d	60. b
10. d	27. d	44. e	61. d
11. e	28. b	45. a	62. b
12. e	29. e	46. e	63. a
13. e	30. d	47. b	64. d
14. d	31. e	48. a	65. d
15. e	32. c	49. a	
16. e	33. e	50. e	
17. e	34. b	51. c	

CHAPTER 3

Cell Division

CELLULAR REPRODUCTION

To create more cells and more living organisms, cells reproduce. However, simply dividing in half is not sufficient. Since it is imperative that each cell pass along its genetic information to succeeding generations of cells, when a cell reproduces, each **daughter cell** must receive more than just a portion of the vital information; a complete copy of all the essential genetic material is necessary.

Some mechanism is needed to pass on all a cell's chromosomal information to its daughter cells. Therefore, before dividing, the parent cell must produce a copy of all the required information. This **duplication** is also called **replication**; both words are used interchangeably when discussing cell division.

Cells, though similar, aren't all identical, and neither are the mechanisms they use to divide. Prokaryotes are thought to represent the most primitive living forms of cells. They divide in a manner that is less complex than the more advanced eukaryotes.

Prokaryotic Cell Division: Binary Fission

Prokaryotic cells have one circular chromosome. When the chromosome duplicates, each of the resulting copies moves to a separate end of the cell, and a membrane forms in the middle of the parent cell, dividing it in two parts. Those two parts then separate, creating two daughter cells; each daughter cell contains an entire set of genetic material. This process of prokaryotic cell division is called **binary fission** (see Figure 3.1).

Eukaryotic Cell Division

Eukaryotic cell division involves a series of steps that are distinct from the process that occurs in prokaryotes. These steps include the division of the nucleus, known as **karyokinesis**, as well as the division of the rest of the cell, which is called **cytokinesis**.

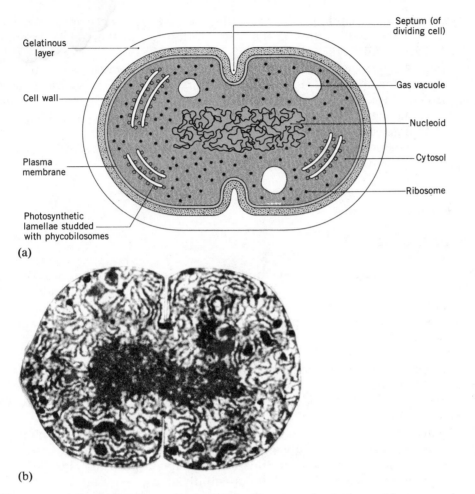

Septum (of
dividing cell)

Gelatinous
layer

Cell wall

Gas vacuole

Nucleoid

Cytosol

Plasma
membrane

Ribosome

Photosynthetic
lamellae studded
with phycobilosomes

(a)

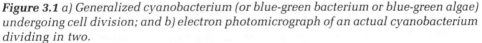

(b)

Figure 3.1 a) *Generalized cyanobacterium (or blue-green bacterium or blue-green algae) undergoing cell division; and b) electron photomicrograph of an actual cyanobacterium dividing in two.*

Before the cell begins to divide, the genetic information located in the nucleus in the form of chromosomes must be duplicated. Only then can the chromosomes be pulled apart, creating two complete sets, one for each incipient daughter cell.

It is only when the two sets of chromosomes are segregated in separate parts of the cell that the cytoplasm and other requisite materials may divide and the parent cell can complete its division.

Interphase

Compared to the rest of the cell's life, cell division is a brief and distinct stage in the cell's life history. **Interphase**, although the longest and most physiologically dynamic part of the cell's life history, is not considered part of cell division. Rather, this is the stage during which the cell is growing, metabolizing, and maintaining itself.

During interphase, the nucleus exists as a distinct organelle, bound by the nuclear membrane. Inside the nucleus are long, thin, unwound strands of chromosomes. While unwound throughout interphase, the chromosomes influence the activities of the cell. It is during interphase that the cell's single set of chromosomes replicates.

Cell division may occur by either **mitosis** or **meiosis**, depending on what type of cell is involved (see Figure 3.2).

MITOSIS

Prophase

Mitosis has begun when the unwound chromosomes begin to coalesce. During their first stage of mitosis, called **prophase**, two small cylindrical bodies become very important, when present (see Figure 3.3). They help organize much of what is about to happen. Located just outside the nucleus in animal cells (though not present in most higher plants), the **centrioles** begin to move to opposite ends of the cell. As this happens, the DNA, which is the primary constituent of the chromosomes, recoils. During this process the chromosomes become more distinct, while the nucleoli become less distinct. It is in the nucleoli that ribosome production occurs; during mitosis, when the chromosomes condense, ribosome production ceases.

When unravelled, the chromosomes interact with their surrounding medium. However, for cell division to occur, the already replicated chromosomes must be pulled apart. Yet this cannot happen until the chromosomes have condensed, making it possible for cell division to proceed.

Once recoiled, the chromosomes become "X" shaped. The X's are composed of two identical **chromatids**, each held to the other by a single **centromere** (see Figure 3.4). One of these chromatids was copied from the other during interphase, when replication occurred.

While the chromosomes are recoiling, the nucleoli and nuclear envelope are disappearing. And at the same time, a series of microtubules are forming the **spindle**. The spindle is the characteristic grouping of microtubules that occurs during nuclear division. It helps to align and move the chromosomes. During the early stages of spindle composition, microtubules radiate around each centriole, creating formations collectively known as the **asters**. Although most higher plants do not have centrioles, they still develop **spindle fibers** at prophase (but asters do not form). The spindle fibers are the microtubules that, together, constitute the mitotic apparatus, called the spindle.

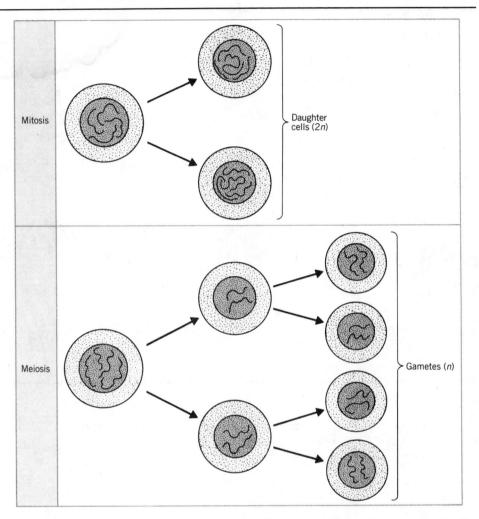

Figure 3.2 *Generalized and somewhat simplified representations of two types of cell division: mitosis and meiosis. In mitosis the two daughter cells have the same number of chromosomes as the parent cell. In meiosis the gametes have half as many chromosomes as the original parent cell.*

During prophase, the chromosomes move toward the middle or the equator of the cell. By the end of prophase, the nuclear membrane is no longer visible; it has broken down.

Metaphase

Metaphase lasts only as long as all the chromosomes remain lined up along the equator. The centrioles have divided in two. Each is attached to one of

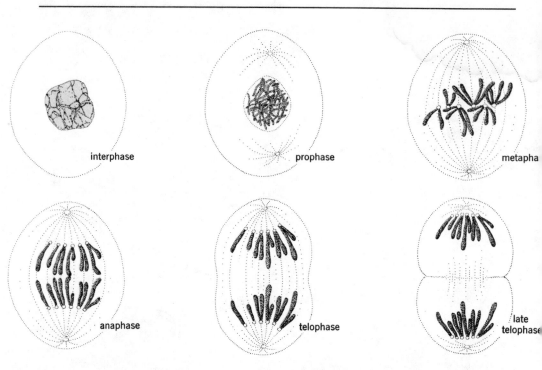

Figure 3.3 *Schematic drawings of an animal cell with six chromosomes undergoing the different stages of mitosis.*

the two corresponding chromosomes from the pair. The individual chromosomes are called **homologs**, and together, both chromosomes in each pair are called **homologous chromosomes**.

Anaphase

Anaphase begins when the two complete sets of chromosomes start moving toward opposite ends of the spindle. Each chromosome appears to be dragged along by its centromere, which is attached to a spindle fiber. The division of the cytoplasm, or **cytokinesis**, begins at the end of anaphase.

Telophase

Telophase is the last stage of mitosis. This is when the cytoplasm separates in two parts of the cell, while the cell's plasma membrane pinches in from both sides, creating two distinct cells. While this is occurring, each set of chromosomes reaches its respective pole, where the nuclear membrane forms, enclosing the chromosomes. The chromosomes then uncoil, while the nucleolus reappears. Upon completion of cytokinesis, a new centriole is made.

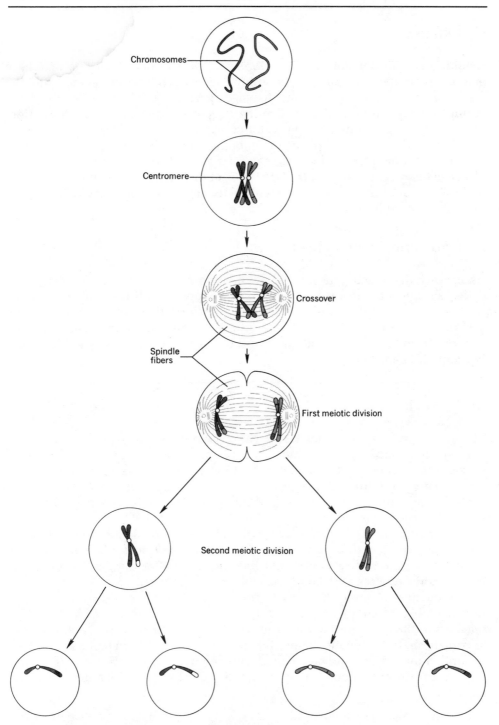

Figure 3.4 *Drawings depicting an animal cell undergoing meiosis. For simplicity and clarity this cell has been shown with one pair of chromosomes. The actual number of chromosomal pairs depends on the species involved. Humans have 23 pairs of chromosomes.*

Cytokinesis

Cytokinesis, the division of the cytoplasm, usually begins in late anaphase and is complete in telophase. In animal cells, cytokinesis begins with the formation of an indentation, or **cleavage furrow**, which forms all the way around the equatorial region, becoming deeper and deeper until it cuts right through, leaving two distinct daughter cells.

Plant cells also have a plasma membrane that divides in the same manner. But in addition, plant cells also have a rigid cell wall that cannot form a cleavage furrow. Instead, a **cell plate** forms at the equatorial region, but instead of moving from the outside in, the cell plate begins forming inside and grows toward the periphery.

Syncytia and Coenocytes

As already discussed, the nuclear membrane reforms during telophase and cytokinesis occurs; however, this does not happen in all types of cells. Some tissues have cells that undergo mitotic divisions that are not followed by cytokinesis, so the cell ends up containing many nuclei. When nuclear division is not followed by cytokinesis, the result in animals is called a **syncytium**, and in plants it is a **coenocyte**.

SEXUALITY

The exchange of genetic material between two cells is a sexual union. With most single-celled organisms, such a sexual union occurs in water. While most prokaryotes reproduce by simple cell division (binary fission), some forms reproduce by **budding**, in which broken off cell fragments grow into mature bacterial cells. Binary fission and budding produce groups of genetically identical cells, known as **clones**.

In some cases, cells exchange or mix their genetic material together, producing populations of unique yet related cells. The transfer is accomplished by a sexual union of two cells that then separate after the genetic exchange occurs. This process of genetic recombination between two cells is known as **conjugation**. Another way that some bacteria exchange genetic material is accomplished simply through the absorption of bits of DNA that were released in the surrounding medium by dead bacteria. This process is known as **transformation**. Genetic material may also be carried from one bacterial cell to another by a virus, a process called **transduction**. Conjugation and transduction occur not only among many bacteria, but also among certain algae and protozoa.

Gametes

Larger organisms that are composed of many cells, called multicellular organisms, can still accomplish sexual union with single cells. In contrast to cells that contain two of each chromosome, sex cells possess only one of each corresponding pair of chromosomes. Such a composition is known as **haploid** or **monoploid**, and it is possible for the two sex cells to unite by forming a single cell, the **zygote**. The merging of genetic material from both cells is called **fertilization**. The zygote has twice the number of chromosomes in each sex cell and is called **diploid**.

Most multicellular organisms have two different types of sex cells, known as **gametes**, such as **eggs** and **sperm**. The sex that produces eggs is female, and the sex producing sperm is male. Sometimes the same individual has both male and female organs. Such an organism is called **monoecious** or **hermaphroditic**. Eggs are usually larger than sperm, since they contain nutrients that help nourish the developing embryo. Unlike eggs, sperm are small and motile (can move on their own).

MEIOSIS

The development of gametes, or **gametogenesis**, occurs through a series of cell divisions known as meiosis. Unlike mitosis, which produces diploid ($2n$) daughter cells, meiosis produces haploid ($1n$) cells, which mature into gametes (sex cells such as sperm and eggs). In contrast to cells that contain two of each chromosome, each sex cell, or gamete, possesses only one of each corresponding pair of chromosomes. This haploid composition makes it possible for the genetic material from both sex cells to unite in fertilization, forming a single cell, the zygote, which is diploid, because once again it has twice the number of chromosomes.

In some respects meiosis is like two back-to-back modified mitotic divisions. For clarity, meiosis has been divided into Meiosis I and Meiosis II. The steps are described in the rest of this chapter and are illustrated in Figure 3.4.

MEIOSIS I

Prophase I

During the first phase of meiosis, **prophase I**, the individual chromosomes coil up and condense, while the **homologous chromosomes** move next to one another. This pairing process is termed **synapsis**. Each chromosome that synapses possesses two chromatids, so that together a series of **tetrads** is formed. Each tetrad consists of four chromatids.

At the time of synapsis, there is an opportunity for genetic material to recombine in new arrangements. This process is called **genetic recombination**. During synapsis, the chromatids may exchange segments of genetic material. When this occurs, it is termed **crossing over**. The recombination of genetic material is a fairly common event. It occurs only during prophase I.

Metaphase I

After prophase I, when the homologous chromosomes have paired up and moved toward the equatorial plane of the spindle, the centromeres line up along the middle, and the centromeres attach to the spindle fibers, each connected to a synaptic pair of chromosomes.

Anaphase I

In **anaphase I**, the centromeres do not divide. Instead, one homolog from each of the homologous pairs moves toward a separate pole.

Telophase I

During **telophase I**, the parent cell splits into two, and the double-stranded chromosomes in the new haploid nuclei fade from view.

Interkinesis

Between mitotic divisions, the genetic material replicates. This does not happen during the brief intervening period after meiosis I and before meiosis II because the chromosomes are already double-stranded. This period is called **interkinesis**.

MEIOSIS II

Prophase II

In meiosis I, the diploid cell produced two haploid cells. During interkinesis, there was no replication of genetic material. In **prophase II**, each chromosome is double-stranded. The chromosomes condense and move toward the equatorial plane, where their centromeres will attach to the spindle fibers.

Metaphase II

Here, the chromosomes line up along the equatorial plane.

Anaphase II

The centromeres then split, and the sister chromatids move toward opposite poles.

Telophase II

In this phase, the chromosomes unwind, the nuclear membranes reform, and the cells divide. Both cells from the beginning of meiosis II were products of a single cell that began at the start of meiosis I. Since both of these cells divided again, the end result of meiosis is that from one cell we get four. Each of the four cells is haploid.

Key Terms

anaphase	interkinesis
anaphase I	interphase
anaphase II	karyokinesis
aster	meiosis
binary fission	metaphase
budding	metaphase I
cell plate	metaphase II
centriole	mitosis
centromere	monoecious
chromatid	monoploid
cleavage furrow	prophase
coenocyte	prophase I
conjugation	prophase II
crossing over	replication
cytokinesis	sperm
daughter cell	spindle
diploid	spindle fiber
duplication	synapsis
egg	syncytium
fertilization	telophase
gamete	telophase I
gametogenesis	telophase II
genetic recombination	tetrads
haploid	transduction
hermaphroditic	transformation
homolog	zygote
homologous chromosomes	

Chapter 3 Self-Test

MULTIPLE-CHOICE QUESTIONS

Cellular Reproduction, Binary Fission, Mitosis, and Meiosis

1. Mitosis and meiosis are both types of _____ .
    a. prokaryotic cell division
    b. eukaryotic cell division
    c. blue-green algae
    d. cytokinesis
    e. binary fission

2. Nuclear division is characterized by chromosome duplication and the formation of two practically identical daughter nuclei known as _____ .
    a. mitosis
    b. meiosis
    c. cytokinesis
    d. chromatin
    e. binary fission

3. Cytokinesis is the division of the _____ .
    a. nucleus
    b. centromere
    c. chromatin
    d. cytoplasm
    e. nucleolus

4. Before a diploid eukaryotic cell begins to divide, the _____ must divide.
    a. nucleus
    b. nuclear membrane
    c. cell wall
    d. chromosomes
    e. buds

5. During most of the life of a eukaryotic cell, it is consuming things, excreting things, growing, and metabolizing. It is only during a brief time when the cell divides. The rest of the time when the cell is not dividing is termed _____ .
    a. interphase
    b. prophase
    c. metaphase
    d. anaphase
    e. telophase

6. Mitosis has begun when two small cylindrical bodies, the ———
   ————— that lie just outside the nucleus, begin to move apart. They are
   present in animal cells, but they are not present in cells of most higher
   plants.

   a. chromosomes
   b. chromatin   C
   c. centrioles
   d. chlorophylls
   d. carotenes

7. During prophase, all the ————————— composing the ———
   ————— coils and condenses into tighter bundles.

   a. DNA, centromeres
   b. DNA, spindle fibers
   c. DNA, chromosomes   e
   d. DNA, asters
   e. RNA, centromeres

8. During mitosis in a diploid cell when the DNA is all wound up, the
   chromosomes can be seen as two long, distinct ————————— .

   a. centromeres
   b. chromatins
   c. asters
   d. chromosomes
   e. chromatids

9. The two identical chromatids are held together by the same ———
   ————— during mitosis in a diploid cell.

   a. chromosome
   b. chromatid
   c. centromere   C
   d. centriole
   e. cell membrane

10. As the centrioles move, each to its opposite pole, a system of thin strands
    of ————————— form around the centrioles in all directions.

    a. fibers
    b. mucus
    c. syncytia
    d. coenocytes
    e. endosperm

11. Some centrioles link up with the fibers from the opposite centriole, and these are called the _____ . The others radiating around each centriole are collectively called the _____ .

    a. chromatids, centromere
    b. spindle fibers, chromatids
    c. asters, spindle fibers
    d. spindle fibers, asters
    e. syncytia, coenocytes

12. During _____ , the chromsomes line up in the middle of the cell.

    a. interphase
    b. prophase
    c. metaphase
    d. anaphase
    e. telophase

13. Metaphase is very brief, lasting only as long as all the chromosomes are attached to their centromeres while lined up along the _____ .

    a. nuclear membrane
    b. cell membrane
    c. opposite poles
    d. equator
    e. coenocytes

14. The moment each centromere divides and they begin to move to opposite poles, each carrying one of the chromatids, metaphase is over and the next phase, which is _____ , has begun.

    a. interphase
    b. prophase
    c. metaphase
    d. anaphase
    e. telophase

15. During anaphase, when the centromeres have split, there are now twice the number of independent _____ in the cell.

    a. centromeres
    b. centrioles
    c. zygotes
    d. spindle fibers
    e. chromosomes

16. During anaphase, each chromosome appears to be dragged along by its
_____ , which is attached to a spindle fiber.

   a. centromere
   b. centriole
   c. cytokinesis
   d. syncytia
   e. coenocytes

17. At the end of anaphase, the division of the cytoplasm, or ____
_____ , begins.

   a. syncytia
   b. coenocytes
   c. synapsis
   d. cytokinesis
   e. interphase

18. During telophase, the following happens:

   a. Cytokinesis is complete, and the nucleolus reappears.
   b. The nuclear membrane forms, and the chromosomes uncoil.
   c. Each set of single-stranded chromsomes is at its respective pole.
   d. Cytokinesis is completed, and the chromosomes uncoil.
   e. All of the above.

19. In animal cells, cytokinesis begins with the formation of an indentation or
_____ that forms all the way around the cell.

   a. cell plate
   b. equatorial plane
   c. endosperm
   d. cleavage furrow
   e. coenocyte

Sexuality and Meiosis

20. When cells come together, exchange genetic material, and then separate,
this is known as _____ .

   a. mitosis
   b. meiosis
   c. binary fission
   d. gametogenesis
   e. conjugation

21. Cells specialized for sexual union are known as _____ .

   a. sex cells
   b. centrioles
   c. bacteria
   d. centromeres
   e. conjugals

22. Most multicellular organisms have two different types of sex cells known
    as _____ .

    a. gonads
    b. testes
    c. ovaries
    d. gametes
    e. eggs

23. Sex cells are produced by a specific series of cell divisions known as
    _____ .

    a. mitosis
    b. binary fission
    c. meiosis
    d. conjugation
    e. gonadogenesis

24. When a nucleus has two of each type of chromosome, the cell is said to
    be _____ .

    a. a gamete
    b. a chromosome
    c. diploid
    d. haploid
    e. polyploid

25. Two haploid cells, known as _____ , unite in fertilization,
    forming a _____ .

    a. zygotes, gamete
    b. zygotes, chromosome
    c. coenocytes, gamete
    d. syncytia, zygote
    e. gametes, zygote

26. Cells that undergo meiosis are _____ .

    a. somatic cells
    b. germ cells
    c. cheek cells
    d. hair follicles
    e. intestine cells

27. Examples of cells that undergo mitosis are _____ .

    a. somatic cells
    b. germ cells
    c. cheek cells
    d. a and b
    e. a and c

28. The pairing of homologous chromosomes during prophase I is called
_____ .

   a. synapsis
   b. metaphase
   c. fission
   d. parthenogenesis
   e. telophase

29. Four _____ are lined up forming a _____ during
synapsis, providing an opportunity for genetic material to recombine in
new ways.

   a. germ cells, genetic recombination
   b. chromatids, tetrad
   c. chromatids, synapsis
   d. tetrad, synapsis
   e. homologous pairs, parthenogenesis

30. During synapsis, it is possible that the chromatids may exchange segments
of genetic material, a process called _____ .

   a. synapsis
   b. parthenogenesis
   c. homology
   d. crossing over
   e. interphase

31. During _____ , homologous chromosomes pair up and move
toward the equatorial plane of the spindle.

   a. mitosis: telophase
   b. mitosis: anaphase
   c. meiosis: prophase I
   d. meiosis: metaphase II
   e. meiosis: anaphase II

32. In mitosis, _____ begins when each centromere carrying its
double-stranded chromosome divides and each single-stranded chromosome starts moving toward the opposite poles of the spindle.

   a. interphase
   b. prophase
   c. metaphase
   d. anaphase
   e. telophase

33. Instead of an interphase, the brief intervening period after meiosis I and
before the commencement of meiosis II is called _____ .

   a. interphase I
   b. interphase II
   c. prophase I
   d. prophase II
   e. interkinesis

ANSWERS

1. b	10. a	19. d	28. a
2. a	11. d	20. e	29. b
3. d	12. c	21. a	30. d
4. d	13. d	22. d	31. c
5. a	14. d	23. c	32. d
6. c	15. e	24. c	33. e
7. e	16. a	25. e	
8. e	17. d	26. b	
9. c	18. e	27. e	

QUESTIONS TO THINK ABOUT

1. Compare and contrast prokaryotic and eukaryotic cell division. Which is less complex and why?
2. In both eukaryotic and prokaryotic cell division, the genetic information must be duplicated before the cell begins to divide. Why?
3. Cell division is a brief and distinct stage in a cell's life history, compared to the rest of a cell's life. Describe what happens during interphase.
4. Define karyokinesis and cytokinesis.
5. Describe, with the use of labeled diagrams, each of the stages in eukaryotic cell division.
6. Define syncytium and coenocyte in terms relating them to cell division.
7. Compare and contrast binary fission, budding, cloning, conjugation, transformation, and transduction.
8. Define haploid and diploid, using the terms *fertilization* and *zygote*.
9. Define the following terms with labeled illustrations, comparing and contrasting the roles these parts play in different stages of a cell's life history: chromosome, chromatid, centromere.
10. With carefully labeled illustrations, describe all the steps in meiosis.

CHAPTER 4

Reproduction

ASEXUAL REPRODUCTION

Asexual reproduction is quite simple compared to sexual reproduction in that it requires only one organism; no partner is necessary. Therefore, in most asexual species, every mature individual can reproduce, enabling the population to increase far more rapidly than the otherwise comparable sexual species that require two individuals to reproduce.

Many organisms take full advantage of asexual reproduction. For instance, each time an *amoeba* divides, it produces two genetically identical replicas of itself. And many species of single-celled as well as multicellular organisms produce asexually reproductive cells known as **spores** that float in the air or water and eventually produce genetic replicas of the parent.

Another asexual mode of reproduction involves **budding**, as illustrated in Figure 4.1. In budding, part of the parent sprouts smaller offspring that separate and become distinct individuals. Many plants reproduce **vegetatively**, sprouting new plants from leaves, roots, or some other part of the parent. In **parthenogenesis**, an egg can develop into an adult without being fertilized by a sperm cell.

Another form of asexual reproduction is **fragmentation**, in which part of an organism separates from the whole, and a new individual regenerates from that part. Such fragmentation sometimes occurs when an organism is in danger; pieces of the injured organism then regenerate into whole organisms. Starfish, for example, have this reproductive capacity. Worms and planaria can also fragment and then **regenerate** the missing portion of their body.

Cloning is another type of asexual reproduction that involves the production of copies that are genetically identical, although they may not look identical. This happens with many plant species—such as when one plant grows from a seed and then many other plants sprout up from the roots.

There are many benefits as well as drawbacks to both asexual and sexual reproduction. Through natural selection, some sexually reproducing organisms have been able to survive and continue to flourish because they benefit from the genetic variability that this mode of reproduction promotes. For instance, should the environment change so that one variation were unable to survive,

Figure 4.1 *Budding is illustrated here, where a parent hydra is "sprouting" a smaller hydra.*

the entire species would perish if the species consisted of a clone. When, however, the species consists of genetically variable individuals within each population, different populations can adapt to a changing or variable environment. Although sexual reproduction is costly and somewhat inefficient in that it takes twice as many individuals to produce one set of offspring, the genetic variability that is maintained through sexual reproduction seems to have long-term benefits.

SEXUAL REPRODUCTION

Sexual reproduction is costly to a species in that it requires both a male and a female to produce as many offspring as one asexual organism can produce, but there are also benefits to such an expensive reproductive mode. The method of reproduction is not always an either/or matter, however; some species reproduce asexually during one part of the year and sexually during another, thereby reaping the benefits of both methods. The complex series of physiological and behavioral changes associated with sexual reproduction are described below.

For sexual reproduction to occur, specialized cells from both the male and female come together and unite. Yet merely combining any two cells is not adequate. Rather, certain cells first undergo a peculiar type of cell division called **meiosis**, creating gametes called **germ cells.**

In animals, undifferentiated male germ cells are located in the **testes**. These cells undergo two meiotic divisions, called **meiosis I** and **meiosis II**, creating four **sperm cells**. Undifferentiated at first, these sperm cells, known as **spermatids**, undergo differentiation before becoming mature **spermatozoa**. This process, **spermatogenesis**, is the result of the division and maturation of a single diploid **primary spermatocyte**, producing four spermatozoa. In female animals, all the undifferentiated germ cells are located in the **ovaries,** where **oogenesis** occurs. Oogenesis comprises the series of steps that produce an **egg** from a **primary oocyte**, which is also called an **ovum**.

Figure 4.2 shows the process and products of meiosis and differentiation for both spermatogenesis and oogenesis. The entire process in which gametes

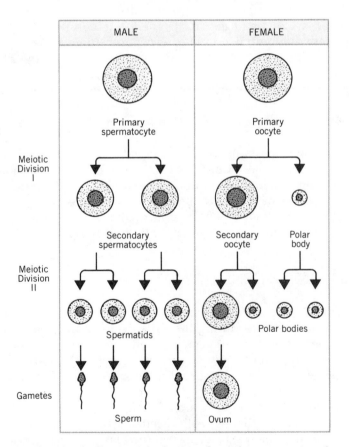

Figure 4.2 *Gametogenesis is illustrated, showing how the process involves meiosis, which, in both the male and female, consists of a first and second meiotic division, forming either sperm or an ovum.*

(sperm and eggs) are developed through meiotic divisions and subsequent maturation and development is known as **gametogenesis**. In the testes of sexually mature male animals, the cells lining the **seminiferous tubules** are always dividing meiotically, producing haploid sperm cells; this process is termed spermatogenesis.

Gametogenesis in females is known as oogenesis, the process that produces eggs in the ovarian structures called **follicles**. The first meiotic division (meiosis I) produces two daughter cells of unequal size from a **primary oocyte**. The primary oocyte divides into a smaller **first polar body** and a larger **secondary oocyte**, which receives a greater share of the cytoplasm during this meiotic division. The first polar body either disintegrates or divides in the second part of meiosis (meiosis II), creating two **second polar bodies** that disintegrate. During meiosis II, the secondary oocyte divides into two parts unequally; less cytoplasm goes to a third second polar body, and more goes toward the **ootid** that then differentiates into an **ovum**, or **egg**.

In terms of weight, a human egg, though extremely tiny (much smaller than the head of a pin), is approximately 58,000 times heavier than a single sperm cell. In terms of length, a completely differentiated sperm cell is about ⅓ the diameter of a human egg. And if lined up side-by-side, there would be seven completely differentiated human eggs per millimeter. With its whip-like motion, the sperm's tail propels it through the mucosal lining of the vagina toward the ovum.

Only one sperm cell can fertilize each egg. When the sperm cell penetrates the egg, it contributes its **haploid** (sometimes called **monoploid** or **1N**) genetic complement of chromosomal DNA to the haploid (1N) egg, creating a **diploid (2N) zygote**.

While spermatogenesis leads to four equal-sized sperm cells because the cytoplasm is divided equally during each meiotic division, oogenesis leads to a single large ovum because of the unequal cytoplasmic divisions. The greater amount of cytoplasm in the egg is used to nourish the embryo. A larger egg may also enhance the chances of being fertilized by a sperm cell. See Figure 4.3 for an illustration of spermatogenesis and oogenesis.

ALTERNATION OF GENERATION

As stated earlier, some organisms use both sexual and asexual reproductive strategies at different stages during their life cycle, a phenomenon termed **alternation of generation**. Such species benefit by possessing the capacity to reproduce asexually, rapidly creating many genetically identical clones, as well as the capacity to reproduce sexually, thus benefiting from genetic recombination.

Cells that develop via meiosis are not always used for reproduction, and the reverse is also true; that is, cells used for reproduction do not always develop via meiosis. For instance, in certain plant species, gametes are produced by mitosis because the cells that produce the gametes are already

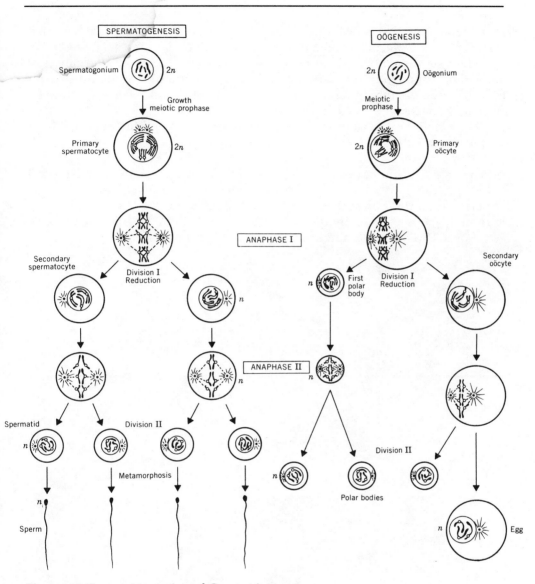

SPERMATOGENESIS

Spermatogonium $2n$

Growth
meiotic prophase

Primary
spermatocyte $2n$

ANAPHASE I

Secondary
spermatocyte

Division I
Reduction n

ANAPHASE II n

Spermatid n

Division II

n

Metamorphosis

n

Sperm

OÖGENESIS

$2n$ Oögonium

Meiotic
prophase

$2n$ Primary
oöcyte

Secondary
oöcyte

n First
polar
body   Division I
Reduction

Division II

n n

Polar bodies

n Egg

Figure 4.3 Spermatogenesis and Oogenesis.

haploid. These haploid gametes then unite in fertilization to form diploid zygotes that divide mitotically, becoming diploid multicellular plants. Eventually, during this stage of their life cycle, haploid cells are produced via meiosis, and the resulting structures are called **spores**. These haploid spores divide mitotically, producing haploid multicellular plants that mature and produce gametes. In some primitive plants, haploid spores were immediately produced by diploid zygotes.

Most variations on the alternation of generation theme are found in the plant kingdom, though there are analogous examples in many animal groups.

For instance, certain coelenterates (cnidarians), or relatives of the jellyfish such as *Hydra*, are sessile, or attached to the substrate. Some species, such as *Obelia*, have both sessile and free-floating forms. This too is a type of alternation of generation, although in this case both life forms are diploid. See Figure 4.4 for an illustration of alternation of generations in *Obelia*, and Figures 4.5 and 4.6 for the life cycles of a moss and a fern.

Abbreviated here is a list of key concepts recounting pertinent points from Chapter 3, on cell division, and from this chapter.

1. When a multicellular, diploid organism produces gametes, the gametes can be a product only of meiosis. If the organism is already haploid as in some plants, then the gametes are the product of mitosis.
2. Haploid cells cannot divide meiotically; only diploid cells can undergo meiosis. Diploid and haploid cells can divide mitotically.

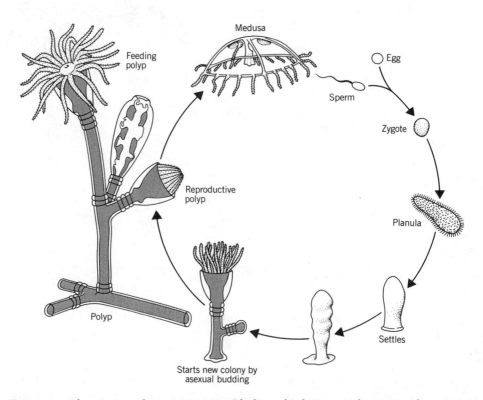

Figure 4.4 *Alternation of generations in Obelia, which is a cnidarian (coelenterate). The polyp is part of the anchored colony (sessile). Reproductive polyps produce medusae, the free-swimming forms that produce eggs and sperm, which combine upon fertilization to form a zygote, which develops into a free-swimming planula larva. The planula eventually settles and through budding forms a new sessile colony.*

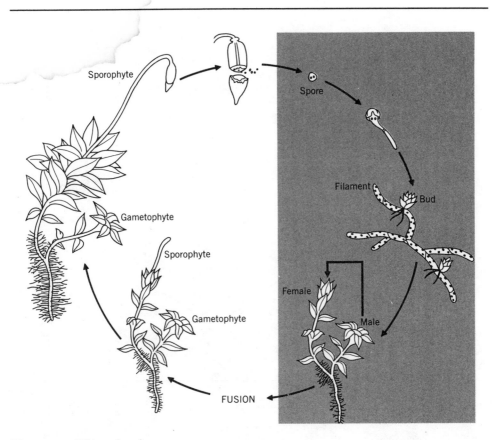

Figure 4.5 Life cycle of a moss.

3. After fertilization, any organism produced by successive mitotic divisions is diploid.
4. If successive mitotic divisions produce a multicellular organism after meiosis, but before fertilization, then the organism is haploid.
5. The diploid phase in the life cycle of most animals is the dominant stage.
6. The haploid stage in the life cycle of most animals represents the stage during which they exist solely as gametes.
7. Animals do not produce spores. Spores are always haploid. To produce a multicellular organism, spores divide mitotically. Gametes unite through sexual reproduction, and fertilization occurs. Later, spores are produced that grow into multicellular organisms through mitotic cell division.
8. In most primitive plants, the haploid stage in the life cycle is dominant.
9. In most higher plants, the diploid stage is dominant.

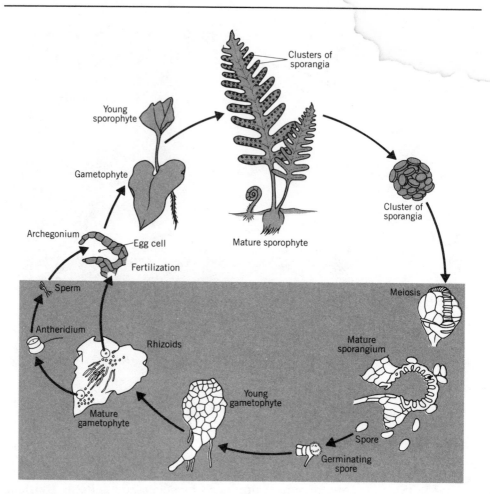

Figure 4.6 *Life cycle of a fern.*

HUMAN REPRODUCTION

The reproductive organs, sexual behavior, and related cycles of humans are similar in many respects to those of other animals, although there are many differences. Here, humans are used as a model to illustrate basic sexual anatomy, physiology, and behavior.

Male Reproductive Organs

As a species that reproduces sexually, humans produce gametes. The male gametes, or sperm, are produced in both **testes** (singular *testis*), which are located within the scrotal sac, or **scrotum**. The testes are composed of tubes, the **seminiferous tubules**, inside of which **spermatogenesis**, or the production of sperm occurs. Spermatogenesis is the series of meiotic divisions that creates the sperm, which in the case of humans is stored in the **epididymis**.

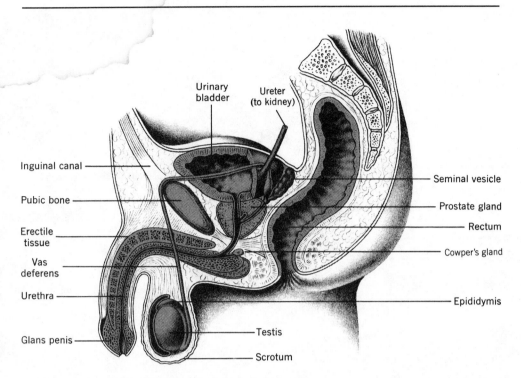

Urinary bladder

Ureter (to kidney)

Inguinal canal

Pubic bone

Erectile tissue

Vas deferens

Urethra

Glans penis

Seminal vesicle

Prostate gland

Rectum

Cowper's gland

Epididymis

Testis

Scrotum

Figure 4.7 *Reproductive system of the human male.*

Upon adequate sexual stimulus, the **penis** becomes erect and sperm moves from the epididymis through the **sperm duct,** where it mixes with secretions from the **seminal vesicles,** the **prostate gland,** and the **Cowper's glands,** producing a mixture known as **semen.** During an orgasm, this fluid is discharged in an **ejaculation.**

Men produce sperm after **puberty,** which may occur any time between the ages of 12 and 17. Puberty usually occurs slightly earlier in girls than in boys. Before puberty, boys can have an erection and are capable of feeling a sensation quite similar to an orgasm. Erections occur even while the male fetus (unborn baby) is in the mother's **womb** (a term that is sometimes used interchangeably with *uterus*). See Figure 4.7 for an illustration of the human male reproductive system.

Female Reproductive Organs

The sex organs outside the body are known as the external genitalia. Together, the external female genitalia are called the **vulva,** which are composed of the protective, sensitive inner and outer **labia** (*labia minora* and *labia majora*). As in the male, urine flows through the **urethra** and out through the opening known as the **urinary meatus.** From back to front on the female, the first opening is the anus, the second is the vagina, and the third is the urinary meatus (see Figure 4.8). Above the urinary meatus is a small structure of which only the protruding tip is exposed. This is the **clitoris,** which swells

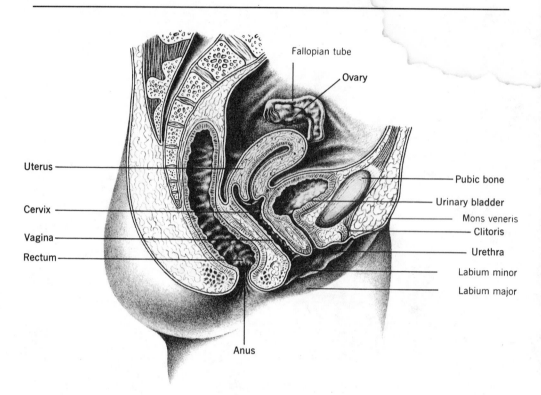

Figure 4.8 *Reproductive system of the human female.*

considerably and hardens when engorged with blood. This is one of the more sensitive parts of the female anatomy, having approximately the same number of nerve endings as a penis. Sensitive to touch, the clitoris and labia are involved in the excitatory responses associated with intimate sexual contact.

Whereas the urethra in the male passes through the center of the penis, the female's urethra opens just above the vaginal orifice. In the female, the urethra connects directly with the bladder, and its sole function is to pass urine. In the male, however, the urethra not only connects with the bladder but during an ejaculation serves as the duct that carries semen.

The **vagina**, sometimes referred to as the birth canal, is the passage leading from the uterus to the vulva. This is where the penis enters during sexual intercourse, and it is through this canal that the baby passes during birth. Girls are born with a thin membrane across the vagina that forms a border around the vaginal opening. This membrane, the **hymen**, has an opening through which such fluids as the menstrual flow pass.

During the foreplay—touching, fondling, and kissing—that precedes inter-course, a significant amount of natural lubrication is secreted by the vagina's **mucosal lining**. Some drops of clear fluid are also secreted through the male's urinary meatus, the hole at the tip of the penis.

Although the vagina may be thought of as an internal reproductive organ, it is in fact a continuation of the exterior of the body. The **cervix**, which is the lower opening of the **uterus**, is the meeting point of the body's exterior and

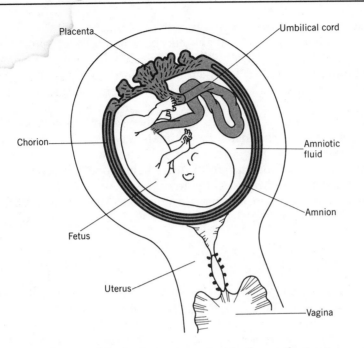

Figure 4.9 Human fetus and its fetal structures inside the mother's uterus.

interior. The uterus is an easily stretched, muscular organ where the embryo develops. When the time comes for the baby to be born, the muscles of the uterus contract, pushing the baby through the cervix and vagina (see Figure 4.9). The cervix is a series of ring-shaped muscles that become dilated or expanded when relaxed, increasing the diameter of the opening. Likewise, when contracted, the opening becomes smaller. This type of muscle is known as a **sphincter**.

The tubes attached to the uterus, also called **oviducts** or **fallopian tubes**, connect the uterus to both of the **ovaries**, which lie on either side of the uterus. When girls are born, each of their immature ovaries already contains all the eggs they will ever have. Some estimates place the number of eggs at 500,000 per ovary, over two-thirds of which die before the female reaches puberty. Most of the rest die during the next several decades until the woman reaches the end of her child-bearing years at **menopause**, which usually occurs at about 45 years of age. During a woman's reproductive years, generally from the age of about 12 or 15 until menopause, one egg is released approximately every 28 days; all together, this means a women releases about 400 to 500 eggs during a lifetime.

Each month, one of the ovaries releases an egg, alternating with its pair from month to month. Following ovulation, the woman is **fertile** for several days, during which time she is capable of becoming pregnant.

When an ovary releases an egg, or **ovulates**, the egg slowly passes down the oviduct aided by the *cilia* toward the uterus. Because the vaginal mucosa are highly acidic, they retard microbial growth. Such acidity would also be injurious to sperm cells without the neutralizing effect of the semen. If **coitus** (sexual intercourse) occurs shortly after ovulation, some of the sperm moving through the vagina pass through the cervix, and move up the uterus to the openings of the oviducts. If a sperm cell reaches the egg and penetrates the egg's membrane, resulting in fertilization, the fertilized egg, or zygote, passes down the oviduct to the uterus, where development continues until the baby is born.

Most other mammals—in fact, most other vertebrates—have sex during specific times. Unlike humans, some animals have mating seasons that occur just once a year. It is during these times that the animals become receptive or stimulated, and the appropriate nuptial behavioral patterns begin. Sexual displays of many types, those specific to each species, are elicited when males and females encounter one another. In contrast, instead of having sex only during specific times in a cycle, humans are receptive all year long.

Menstrual Cycle

Puberty is the point at which an individual is first capable of reproduction. Girls usually reach this stage when about 12 years old, and boys reach puberty about two years later. When a girl reaches puberty, she has her first menstrual period, and her periods recur every 28 days or so during the rest of her child-bearing years. It is the female monthly cycle that is known as the menstrual cycle.

Blood and cells that had lined the uterus are expelled in the menstrual flow, which is a result of the breakdown of the uterine lining. The outer cells slough off in a readying process that allows the uterus to regenerate a new lining receptive to implantation of a fertilized egg, creating a proper environment where an embryo can develop to full term. During estrus, when the cells are sloughed off, old cells and accompanying blood drain from the uterus through the vagina, out of which the discharge flows. This bleeding is controlled and normal, and is referred to as a woman's period. In humans, the entire menstrual cycle of uterine buildup and breakdown, involving the release of one egg, takes 28 days. The term, menstrual cycle, originated from the root menses, which means month. Menses is also used when referring to the menstrual period.

Many changes occur in hormonal concentrations during the menstrual cycle. Two parts of the monthly cycle are commonly known as the ovarian cycle, and the uterine cycle. The relative levels of estrogen and progesterone are two important hormones that are involved in regulating the events of the monthly cycle. Figure 4.10 explains the events in more detail.

The follicle is the part of the ovary that releases the egg. While the follicle develops, it secretes estrogen, which stimulates the uterus to thicken. Ovulation typically occurs on the fourteenth day of the cycle (counting begins from the first day of flow). The ruptured follicle becomes the corpus luteum, which secretes another hormone, progesterone, which stimulates the uterus to become ready to receive the fertilized egg. If the egg does not become fertilized, then progesterone secretion drops off, the thickened lining of the

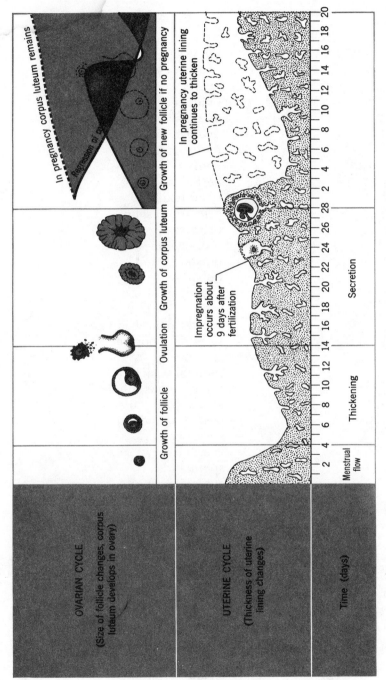

Figure 4.10 *Monthly ovarian and uterine cycles. Four hormones are involved: FSH (follicle stimulating hormone) causes the maturity of the egg; Estradiol causes the uterine lining to grow thicker; LH (luteinizing hormone) causes ovulation; and progesterone helps prepare the lining of the uterus for the fertilized egg. This is the "ovarian" cycle. If the egg is not fertilized, it passes through the uterus and disintegrates, the corpus luteum stops producing progesterone, and menses occurs. This is the "uterine" cycle.*

uterus sloughs off, and bleeding signals the beginning of the next menstrual cycle (see Figure 4.10).

Key Terms

alternation of generation
asexual reproduction
budding
cervix
cilia
clitoris
cloning
coitus
corpus luteum
Cowper's gland
diploid
egg
ejaculation
epididymis
estrogen
estrus
fallopian tube
fertile
first polar body
follicle
fragmentation
gametogenesis
germ cells
haploid
hymen
inner labia
labia
labia majora
labia minora
meiosis
meiosis I
meiosis II
menopause
menses
menstrual cycle
menstrual period
monoploid
mucosal lining
1N
2N
oogenesis
ootid

outer labia
ovarian cycle
ovary
oviduct
ovulate
ovum
parthenogenesis
penis
period
polar bodies
primary oocyte
primary spermatocyte
progesterone
prostate gland
puberty
regeneration
scrotum
second polar bodies
secondary oocyte
semen
seminal vesicle
seminiferous tubules
sexual intercourse
sexual reproduction
sperm cells
sperm duct
spermatids
spermatogenesis
spermatozoa
sphincter
spores
testes
urethra
urinary meatus
uterine cycle
uterus
vagina
vegetative reproduction
vulva
womb
zygote

Chapter 4 Self-Test

QUESTIONS TO THINK ABOUT

1. Define asexual reproduction and discuss the different types that exist.
2. Compare and contrast asexual and sexual reproduction: What are the benefits and shortcomings of each?
3. Describe with the use of illustrations both types of gametogenesis: spermatogenesis and oogenesis.
4. List the main human male reproductive structures (internal and external), and explain their function.
5. List the main female reproductive structures (internal and external), and explain their function.
6. Give the precise route that sperm takes from where it is formed to where it leaves the body.
7. Give the precise route that an egg takes from where it is formed to where it becomes an embryo, and then the path it takes when the baby is born.
8. Describe a complete menstrual cycle.

MULTIPLE-CHOICE QUESTIONS

Asexual Reproduction

1. Is **cloning** a form of *sexual* or *asexual* reproduction?
2. Is **parthenogenesis** a type of asexual reproduction where an egg *can* or *cannot* develop into an adult without being fertilized by a sperm cell?
3. Many plants reproduce _____ , sprouting new plants from leaves or roots or some other part of the parent.
    a. by budding
    b. vegetatively
    c. parthenogenetically
    d. altruistically
    e. parsimoniously
4. Another form of asexual reproduction involves _____ , which occurs when part of an organism separates from the whole, and from this piece a new individual regenerates.
    a. fragmentation
    b. parthenogenesis
    c. parsimony
    d. altruism
    e. alternation of generation

5. In the testes of sexually mature male animals, the cells lining the
_____ divide meiotically, producing haploid sperm cells.

    a. ovaries
    b. kidneys
    c. seminiferous tubules
    d. ureters
    e. urethras

6. Gametogenesis in female animals is known as _____ .

    a. follicular growth
    b. spermatogenesis
    c. oogenesis
    d. mitosis I
    e. mitosis II

7. Some organisms use both sexual and asexual reproductive strategies at
different stages during their life cycle, which is termed _____ .

    a. sporation
    b. ovulation
    c. sex
    d. alternation of generation
    e. spermatogenesis

8. When a multicellular, diploid organism produces gametes, the gametes
can only be a product of _____ .

    a. meiosis
    b. mitosis
    c. alternation of generation
    d. spermatogenesis
    e. oogenesis

9. Animals _____ produce spores.

    a. sometimes
    b. always
    c. never
    d. meiotically
    e. mitotically

10. The _____ phase in the life cycle of most animals is the dominant stage.

    a. monoploid
    b. haploid
    c. diploid
    d. triploid
    e. tetraploid

Sexual Reproduction

11. In spermatogenesis, from one primary spermatocyte develop _____
    _____ .

    a. two secondary spermatocytes
    b. four secondary spermatocytes
    c. two spermatids
    d. four spermatids
    e. a and d

12. Upon completion of meiosis I and meiosis II in oogenesis, the primary
    oocyte has developed into _____ .

    a. one secondary oocyte, one ootid, and one ovum
    b. one first polar body and three second polar bodies
    c. one secondary oocyte and one first polar body
    d. one ootid and three second polar bodies
    e. one ovum and three polar bodies

13. When the genetic material in the _____ unite, the union is
    known as fertilization.

    a. gametes
    b. ootids
    c. secondary polar bodies
    d. spermatocytes
    e. oocytes

14. In male vertebrates, the sperm cells are formed by the cells lining the
    _____ .

    a. follicles
    b. spores
    c. zygotes
    d. seminiferous tubules
    e. mesothelium

15. In the testes, the epithelial cells divide meiotically, producing _____
    _____ sperm cells.

    a. haploid
    b. diploid
    c. tetraploid
    d. haplodiploid
    e. all of the above

16. The aggregation of cells in the ovary that releases the egg is known as a
    _____ .

    a. menses
    b. follicle
    c. progesterone
    d. estrogen
    e. period

17. Many types of plants and animals reproduce by a strategy that uses both sexual and asexual reproduction, known as _____ .
    a. parthenogenesis
    b. sporangia
    c. oogenesis
    d. genetic recombination
    e. alternation of generation

18. The haploid cells that plants sometimes produce meiotically are called _____ .

    a. oocytes
    b. spermatocytes
    c. seeds
    d. spores
    e. zygotes

19. An asexual mode of reproduction known as _____ occurs when part of the parent sprouts smaller offspring that then become separate individuals.
    a. budding
    b. sprouting
    c. germinating
    d. spreading
    e. parthenogenerating

20. A form of asexual reproduction that produces genetically identical offspring is known as _____ .
    a. germination
    b. gestation
    c. fragmentation
    d. cloning
    e. parthenogenesis

Human Reproduction

21. The sac in which the testes are located is the _____ .
    a. seminal vesicle
    b. prostate
    c. epididymis
    d. scrotum
    e. Cowper's sac

22. The production of sperm occurs inside the _____ .
    a. epididymis
    b. seminal vesicles
    c. prostate gland
    d. seminiferous tubules
    e. Cowper's gland

23. Toward the top of each testis are some coiled tubes inside which sperm is stored; this area is known as the ——————— .

    a. seminal vesicle
    b. sperm duct
    c. prostate gland
    d. seminiferous tubules
    e. epididymis

24. When a male ejaculates, sperm moves from the epididymis through the sperm duct and is then mixed with secretions from the ——————— and ——————— .

    a. prostate gland, thymus gland
    b. vagina, seminal vesicle
    c. urethra, vulva
    d. Cowper's gland, urinary meatus
    e. seminal vesicle, prostate gland

25. When sperm moves through the sperm duct and is mixed with the above secretions, it becomes known as ——————— .

    a. mucous
    b. spermicidal jelly
    c. prostoglandin
    d. semen
    e. urine

26. The ——————— passes through the length of the center of the penis.

    a. ureter
    b. urethra
    c. urinary meatus
    d. vulva
    e. vagina

27. The passage leading from the uterus to the vulva is known as the ——————— .

    a. ureter
    b. urethra
    c. urinary meatus
    d. vulva
    e. vagina

28. The ——————— is a muscular organ where the embryo develops.

    a. cervix
    b. sphincter
    c. uterus
    d. ovary
    e. menopause

29. The lower part of the uterus that extends into the vagina is the _____
    _____ .

   a. ovary
   b. cervix
   c. oviduct
   d. uvula
   e. pubis

ANSWERS

1. asexual	9. c	17. e	25. d
2. can	10. c	18. d	26. b
3. b	11. e	19. a	27. e
4. a	12. e	20. d	28. c
5. c	13. a	21. d	29. b
6. c	14. d	22. d	
7. d	15. a	23. e	
8. a	16. b	24. e	

CHAPTER 5

Cellular Respiration

Cellular respiration is a series of chemical reactions that frees the energy in fat, protein, and carbohydrate food molecules, rendering it available to the cells (see Figure 5.1). Respiration is generally defined as the oxygen-requiring stage in these biochemical reactions. However, in certain instances, respiration also occurs without any oxygen; this is known as **anaerobic respiration**.

During respiration, as in photosynthesis (described in Chapter 6), each chemical reaction is catalyzed by an enzyme. To break down glucose molecules, which are the stable end products of photosynthesis, ATP is needed to provide the **activation energy** to initiate the chemical processes that follow. ATP is one of the major energy-providing molecules that initiate biochemical reactions throughout the body. Because ATP, $NADH_2$, and similar molecules are essential to the maintenance of living systems, organisms need to ensure the constant supply of such energy sources. See Figure 5.2 for the role of enzymes in reducing activation energy and Figure 5.3 for the role of temperature in enzyme activity.

GLYCOLYSIS

Glycolysis is the first series of chemical reactions in cellular respiration, in which glucose is converted to pyruvate (pyruvic acid) and, depending on the organism and the particular conditions involved, pyruvate is processed further into other critical end products (see Figure 5.4).

Oxidation, chemically defined as the loss of an electron, usually takes place through the addition of oxygen or the subtraction of hydrogen. The oxidation of glucose to pyruvate releases energy, most of which would be lost as heat if it were not conserved. To conserve the released energy so it may be harnessed to run the cell's metabolism, the cell regulates its chemical reactions. Chemical reactions break down glucose in controlled incremental steps. Together, these steps are called glycolyis (meaning the *lysis*, or splitting, of glucose). Glucose, a 6-carbon sugar, is first split into two 3-carbon fragments, one of which is pyruvate. The other fragment is then converted to pyruvate.

Each of the intermediary steps involved in converting glucose to pyruvate occurs within the cell's cytoplasm. The entire process can take place whether

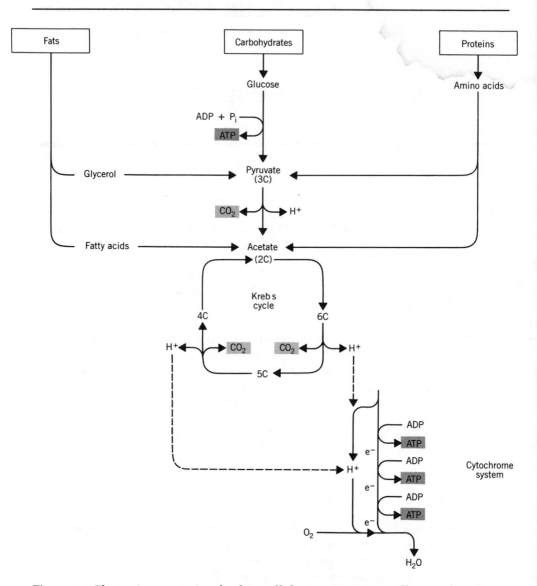

Figure 5.1 *The major events involved in cellular respiration are illustrated in this flow diagram. First carbohydrates, fats, and proteins are digested into pyruvate and acetate molecules, producing ATP. The acetate molecules enter the Krebs cycle, where carbon dioxide (CO_2), hydrogen ions (H^+), and electrons (e^-) are produced. The energy from the hydrogen ions, which are protons, and the electrons is used to create ATP molecules in the cytochrome system. Then the hydrogen ions combine with the electrons to form water.*

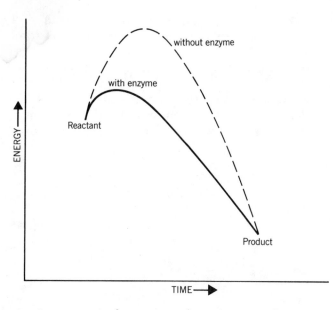

Figure 5.2 *Activation energy is the amount of energy required to initiate a chemical reaction; enzymes can reduce the amount of activation energy that is necessary.*

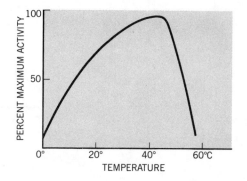

Figure 5.3 *Different enzymes function best at specific temperatures. Most human enzymes function best at the normal body temperature of 37°C (= 98.6°F).*

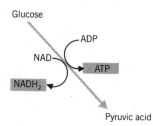

Figure 5.4 *This figure illustrates the most important points of what occurs during the several steps of biochemical reactions, known as glycolysis, which converts glucose to pyruvic acid (CH_3–CO–COOH).*

or not oxygen is present, because no molecular oxygen is used. This explains why glycolysis is sometimes referred to as **anaerobic respiration**. These nine steps are described and illustrated in Figure 5.5.

In the first step of glycolysis, one of the phosphate groups from ATP is added to glucose (see Figure 5.6 for an illustration of ATP's structure). Then the glucose, with a phosphate group on the sixth carbon (hence glucose 6-phosphate) undergoes a rearrangement, retaining the same number of atoms and becoming another 6-carbon compound, fructose 6-phosphate. At this

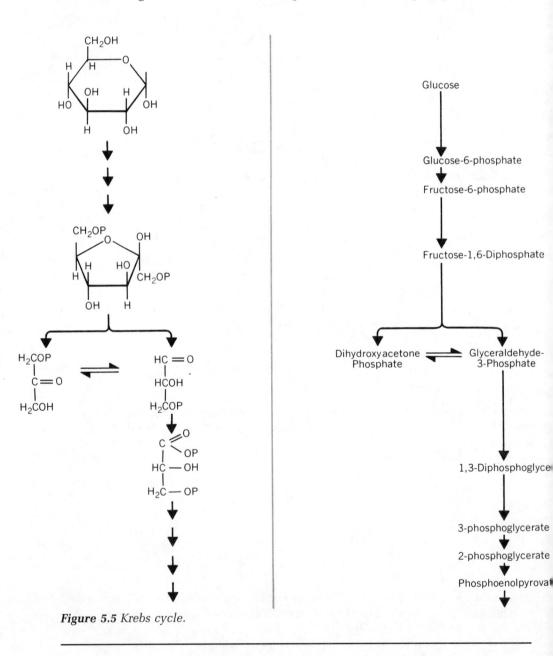

Figure 5.5 *Krebs cycle.*

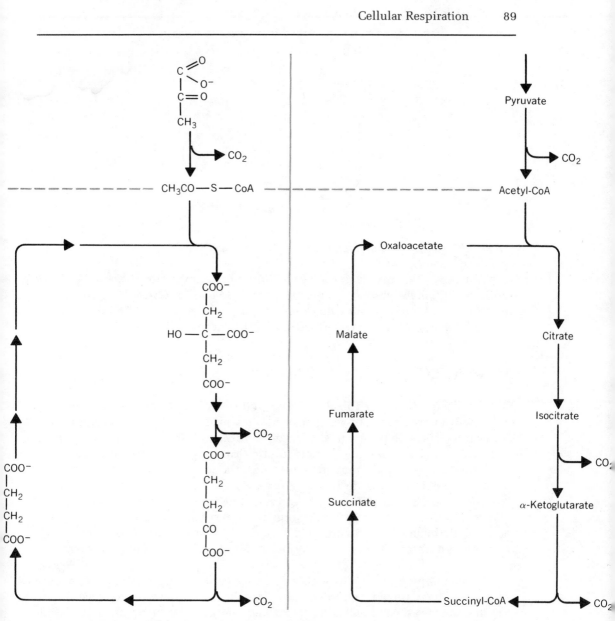

Figure 5.5 (continued) *Krebs cycle and glycolysis.*

point, another ATP donates a phosphate group, and the previous compound becomes fructose 1,6-diphosphate, which is then split into two 3-carbon molecules, each having a phosphate group. One of these 3-carbon compounds, phosphoglyceraldehyde (PGAL), was an endproduct of photosynthesis before being converted into glucose. The other 3-carbon compound that resulted from splitting fructose biphosphate is converted into PGAL. So far, the glycolytic pathway has consumed two ATPs to convert glucose back into PGAL, which brings us back to the second-to-last step of photosynthesis, before the PGALs were converted into glucose.

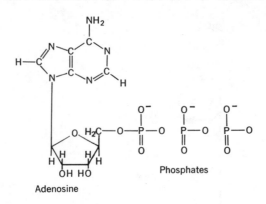

Figure 5.6 *The structure of adenosine triphosphate (ATP).*

The next step involves oxidation and phosphorylation (the addition of another phosphate group), converting the PGAL into 1,3-diphosphoglyceric acid. One high-energy phosphate group on each of the diphosphoglycerate molecules is then transferred to ADP, making more ATP molecules.

ANAEROBIC FERMENTATION

Many cells, particularly those in most plants and in some microorganisms, can obtain energy without oxygen by the anaerobic process of **fermentation**. Without oxygen, the pyruvate will accept hydrogen from NADH$_2$, freeing more NAD to accept hydrogen in other glycolytic reactions. When added to the pyruvate, the hydrogen will convert it into ethyl alcohol, a 2-carbon compound, and release carbon dioxide in the process. This anaerobic pathway converts pyruvate into alcohol (ethanol). In animals and some microorganisms, anaerobic respiration reduces (adds hydrogen to) pyruvate, producing energy and lactic acid (lactate is a 3-carbon compound). The anaerobic processes that form ethanol or lactate are called fermentation. For an illustration of them, see Figure 5.7.

If the energy released through a series of controlled glycolytic steps were released at once, the heat could cause a fire. This explains why cells use small, incremental steps, each releasing controlled energy either to synthesize ATP or to help attach hydrogen atoms to molecules of the hydrogen-carrying coenzyme NAD$^+$ (**nicotinamide adenine dinucleotide**) to form NADH$_2$.

The enzymes that engage each glycolytic step are dissolved in the cytoplasm. For other carbohydrates to enter the glycolytic pathway, they too must first be converted to glucose. Enzymes then remove water molecules and restructure the **phosphoglycerate** molecules into **phosphoenolpyruvate**, which has a high-energy bond connecting the phosphate group. This last phosphate group is then used to phosphorylate ADP to ATP, converting phosphoenolpyruvate into the 3-carbon pyruvate.

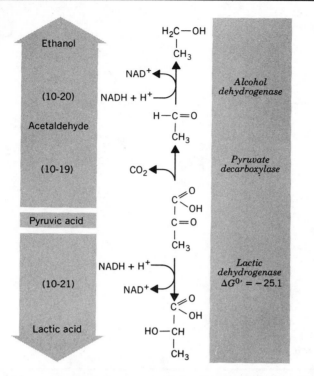

Figure 5.7 *Anaerobic conversion of pyruvate (pyruvic acid) to ethanol and to lactate (lactic acid). This process is also called anaerobic respiration and alcoholic fermentation.*

AEROBIC RESPIRATION

In the presence of oxygen, the 3-carbon pyruvate molecules produced in glycolysis can be further oxidized. The aerobic respiration of pyruvate, which occurs within the mitochondria, takes place in two stages: the Krebs cycle (named for Hans Krebs of England, who discovered it in 1953) and oxidative phosphorylation.

Organisms that use glycolysis with respiration produce 19 times as much energy as do anaerobic species breaking down the same food. Before this efficient metabolic process evolved, there was anaerobic glycolysis. The development of the Krebs cycle (illustrated in Figure 5.8) increased the number of potential metabolic pathways and raised the number of ATPs produced to 38. The Krebs cycle, part of aerobic cellular respiration, is the most common pathway for the oxidative metabolism of pyruvate. After pyruvate is converted to acetate, coenzyme A (CoA) is attached to it, forming acetyl-CoA, which enters the Krebs cycle.

CYTOCHROME TRANSPORT SYSTEM

When the carbon atoms of glucose molecules are oxidized, some of the resulting energy is used to add a phosphate to adenosine diphosphate (ADP), creat-

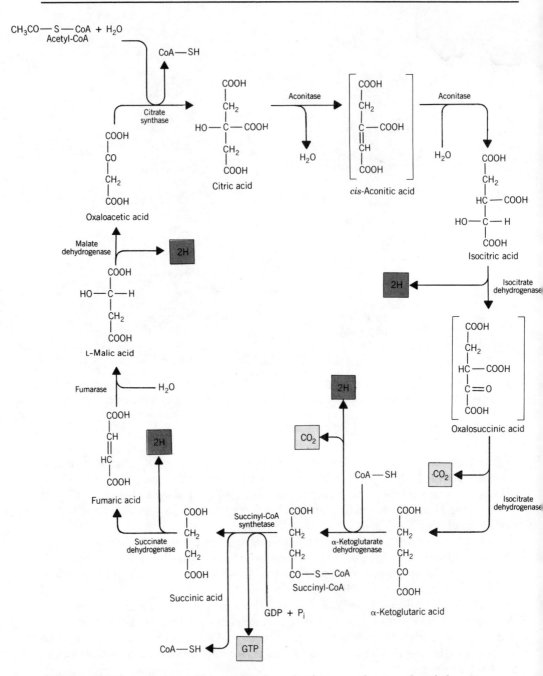

Figure 5.8 *The Krebs (tricarboxylic acid) cycle shown with more detail than in Figure 5.5.*

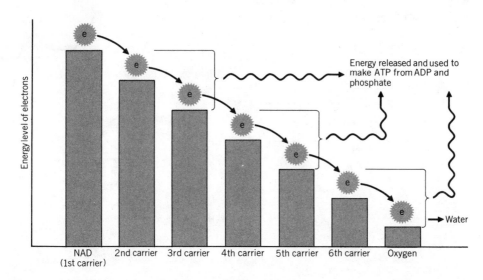

Figure 5.9 *Oxidative phosphorylation is the net product of the cytochrome system, which is an electron transport chain that gradually releases energy during a series of steps via iron-containing molecules called cytochromes. This energy, which is stored in the stable housing of bonds formed when phosphorus combines with ADP, producing ATP, is used later to run the many energy-requiring processes within each cell.*

ing adenosine triphosphate (ATP). Most of the energy, however, remains in the high-energy carbon-hydrogen bonds and in the high-energy electron carriers NAD^+ and FAD. In the final stage of the oxidation of glucose, the high-energy-level electrons are passed sequentially in a controlled, step-by-step process, to lower-energy-level electron carriers. A small amount of energy is released during each step that is used to form ATP from ADP. This phase of aerobic respiration is known as **oxidative phosphorylation** (see Figure 5.9).

The NADH and $FADH_2$ released during the Krebs cycle are transported to the series of electron carriers known as the **electron transport chain**. The main components of the electron transport chain are **cytochromes** (iron-containing protein molecules). At the end of the chain, the electrons are accepted by oxygen atoms, producing water.

In a simplified manner, aerobic respiration may be summarized in this formula:

$$C_6H_{12}O_6 + 6O_2 \rightarrow 6CO_2 + 6H_2O + energy$$

The steps represented in this equation occur in a coordinated process. These are illustrated in Figure 5.10.

RESPIRATION OF FATS AND PROTEINS

As stated earlier, more than just carbohydrates can be fed into the respiratory chain of reactions. Other molecules may also be modified and metabolized.

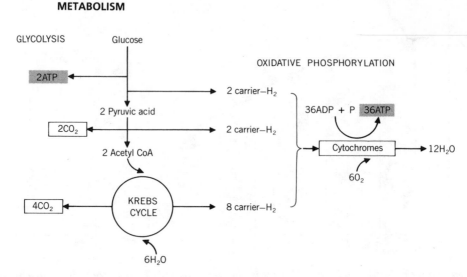

Figure 5.10 *Summary diagram of the formation of ATP during the aerobic breakdown of glucose to carbon dioxide and water. The integration of the three processes— glycolysis, the Krebs cycle, and oxidative phosphorylation—is illustrated.*

The first step in breaking down fats to produce glycerol and fatty acids is **hydrolysis**, in which the molecules are broken down through the addition of water. Glycerol, which is a 3-carbon compound, is converted into PGAL and fed into the glycolytic pathway. The fatty acids are then further broken down into carbon fragments, which are converted into acetyl-CoA and then fed into the respiratory pathway. When metabolized, gram for gram, fat yields a little more than twice as much energy as carbohydrates.

The hydrolysis of protein produces amino acids that may then be metabolized by several different means. One is **deamination**, or the removal of the amino group. In another method, amino acids are converted into pyruvic acid. Through other mechanisms, amino acids are converted into acetyl-CoA, while others are converted into compounds that can be used in the Krebs cycle. Gram for gram, the complete oxidation of protein yields about the same amount of energy as carbohydrates.

Key Terms

activation energy
aerobic respiration
anaerobic respiration
cellular respiration
cytochrome transport system
deamination
electron transport chain

fermentation
glycolysis
hydrolysis
Krebs cycle
mitochrondia
oxidation
oxidativé phosphorylation

Chapter 5 Self-Test

QUESTIONS TO THINK ABOUT

1. Under what circumstances might cellular respiration occur in the presence of oxygen, and when might oxygen be unnecessary?
2. Although it has been stated that the oxidation of glucose to pyruvate releases energy, what exactly is meant by the term oxidation?
3. The first series of chemical reactions in cellular respiration, in which glucose is converted to pyruvate, is called glycolysis. Expand on what happens during this series of reactions.
4. Why are enzymes important in glycolysis?
5. How are both glycolysis and the Krebs cycle related?
6. Describe anaerobic fermentation; compare and contrast it to other ways cells have to obtain energy. Integrate anaerobic fermentation into the larger picture, explaining the role it plays in most plants and in some microorganisms.
7. How do ATP and ADP differ, and why are they important in cellular respiration?
8. What happens during the Krebs cycle?
9. What is the electron transport chain, what role do the cytochromes play in this chain, and where might such an electron transport chain exist? Why is it so important?
10. What types of substances can be fed into the respiratory chain of reactions? And what happens to them once they have been fed into this chain of reactions?

MULTIPLE-CHOICE QUESTIONS

1. Fats, proteins, and carbohydrates are reservoirs of energy that through a series of chemical reactions known as _____ can become available to the cells.

   a. respiration
   b. oxidation
   c. glycolysis
   d. transformation
   3. activation

2. In eukaryotes, all the respiratory steps except for glycolyis occur within the

   _____ .

   a. nucleus
   b. ribosomes
   c. golgi bodies
   d. lysosomes
   e. mitochondria

3. Glycolysis occurs in the _____ .

   a. nucleus
   b. nucleoplasm
   c. nucleolus
   d. cytoplasm
   e. a, b, and c

4. Gram for gram, the complete oxidation of protein yields _____ energy as that of carbohydrates.

   a. about the same amount of
   b. about twice as much
   c. about three times as much
   d. one-half the amount of
   e. one-quarter the amount of

5. Many cells, particularly those in plants and in some microorganisms, can obtain energy without oxygen by the anaerobic process of _____ .

   a. fermentation
   b. aerobic respiration
   c. deamination
   d. deoxidation
   e. transformation

6. The aerobic respiration of pyruvate occurs within the _____ .

   a. nucleus
   b. cytoplasm
   c. mitochondria
   d. golgi bodies
   e. lysosomes

7. With the aid of a complex of enzymes found in the mitochondrial matrix, acetate is broken down through a long series of chemical reactions known as (the) _____ .

   a. Krebs cycle
   b. osmosis
   c. deamination
   d. oxidation
   e. mitochondrial cycle

8. In the final stage of the oxidation of glucose, high energy electrons are passed in a controlled series of steps to lower energy level electron carriers, with a small amount of energy released during each step; the energy is used to form ATP. This phase of aerobic respiration is known as _____ _____ .

   a. oxidative phosphorylation
   b. Krebs cycle
   c. citric acid cycle
   d. tricarboxylic acid cycle
   e. anaerobic fermentation

9. In the _____ , the 3-carbon pyruvic acid molecules produced through glycolysis can be further oxidized.

   a. presence of oxygen
   b. absence of oxygen
   c. nucleus
   d. lysosome
   e. ribonucleic acid

10. Pyruvate is passed to the _____ , where in eukaryotic cells all the respiratory steps except for glycolysis occur.

   a. lysosomes
   b. infundibulum
   c. nucleus
   d. Golgi apparatus
   e. mitochondria

11. Pyruvate is transported to the mitochondrial matrix where the 3-carbon compound is converted into the 2-carbon acetate. Then, with the aid of the complex of enzymes found in the mitochondrial matrix, the acetate begins the long series of chemical reactions known as _____ .

   a. the Krebs cycle
   b. citric acid cycle
   c. tricarboxylic acid cycle
   d. all of the above
   e. none of the above

12. During the Krebs cycle, the most important by-products are _____ _____ and several hydrogen-carrier molecules including _____ _____ and some _____ .

   a. ATP
   b. NADH
   c. $FADH_2$
   d. all of the above
   e. none of the above

13. Organisms that use glycolysis with respiration produce ————————
    as anaerobic species breaking down the same food.
    a. energy just as efficiently
    b. energy less efficiently
    c. energy twice as efficiently
    d. energy three times as efficiently
    e. energy 19 times as efficiently

14. In the final stage of the oxidation of glucose, the high energy level elec-
    trons are passed sequentially in a controlled, step-by-step process to lower-
    energy-level electron carriers. A small amount of energy is released dur-
    ing each step and is used to form ATP and ADP. This phase of aerobic
    respiration is known as ——————————.
    a. the Krebs cycle
    b. the tricarboxylic cycle
    c. the citric acid cycle
    d. oxidative phosphorylation
    e. all of the above

15. The NADH and $FADH_2$ released during the Krebs cycle are transported to
    the series of electron carriers in the phase of anaerobic respiration known
    as oxidative phosphorylation. This series of electron carriers is known as

    ————————.
    a. the electron transport chain
    b. the tricarboxylic cycle
    c. citric acid cycle
    d. amination
    e. deamination

16. The hydrolysis of proteins through the removal of the amino group is one
    of several ways that proteins may be metabolized, producing amino acids;
    this method is known as ——————————.
    a. amination
    b. deamination
    c. carbolysis
    d. glycolysis
    e. transformation

ANSWERS

1. a	5. a	9. a	13. d
2. e	6. c	10. e	14. d
3. d	7. a	11. d	15. a
4. a	8. a	12. d	16. b

CHAPTER 6

Photosynthesis

Every leaf is essentially a solar-powered carbohydrate factory where, fueled by the sun, raw materials such as carbon dioxide and water are transformed by the complex molecular machinery into stable, energy-rich, finished products that help run nature's entire economy. This solar-powered process that makes carbohydrates is known as **photosynthesis**. It occurs in the leaves of higher plants, as well as in many other plant parts, especially in those that are green. Photosynthesis also occurs in a range of organisms other than plants, including some bacteria and protists.

HISTORY

In 1772, Joseph Priestly, a British clergyman and chemist, demonstrated that when a plant or animal was kept alone in an airtight jar, it died. However, when a plant and an animal were put together in a jar, both lived. Seven years later the Dutch physician Jan Ingen-Housz showed that sunlight was necessary for plants to produce oxygen although, like Priestly, he knew nothing about oxygen at the time and explained his results in another way. Then in 1782, a Swiss pastor and part-time scientist, Jean Senebier, showed that plants use carbon dioxide (CO_2) when they produce oxygen (O_2). He suggested that CO_2 was converted to O_2 during photosynthesis.

Again, it should be stressed that Senebier didn't know which gases were involved. Rather, he reported that the process was dependent upon a particular kind of gas, which he called "fixed air," and which we now know as carbon dioxide. Then in 1804, the Swiss worker Nicolas Theodore de Saussure found that water is necessary for the photosynthetic production of organic materials. So by the early nineteenth century, the basic ingredients involved in photosynthesis were already known and could be put in the following equation (see also Figure 6.1).

$$\text{carbon dioxide} + \text{water} \xrightarrow[\text{light}]{\text{green plants}} \text{organic material} + \text{oxygen}$$

In 1883 T. W. Engelmann, a German researcher, conducted an experiment that provided circumstantial evidence indicating that **chlorophyll**, the green

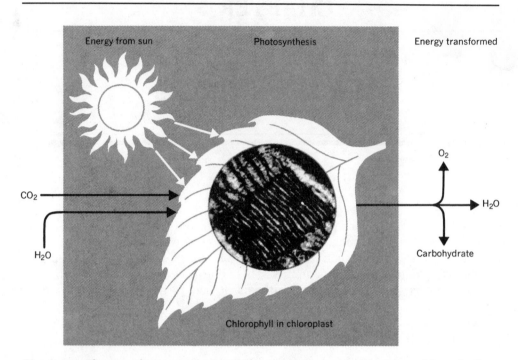

Figure 6.1 *Photosynthesis, as depicted here, is the process involving chlorophyll molecules that use the energy in sunlight to convert carbon dioxide (CO_2) and water (H_2O) into carbohydrate.*

pigment of plants, might be important in photosynthesis. He studied the algae *Spirogyra sp.*, which has distinctively long, spiral chloroplasts. With a prism, Engelmann directed specific wavelengths of light to different parts of the *Spirogyra*. He then introduced oxygen-requiring bacteria to the solution, expecting that when the *Spirogyra* was getting the best light for photosynthesis, the *Spirogyra* would release the most oxygen, and that would be where the bacteria would move, for the oxygen. The greatest numbers of bacteria clustered where the chloroplasts were absorbing the bands of red and blue light, not by the green wavelengths. From this, Engelmann deduced that oxygen was being produced where the red and blue wavelengths were being absorbed.

PHOTOSYNTHETIC PIGMENTS

The reason most leaves look green is because while they absorb the light's red and blue wavelengths, the green passes through, so that's what we see. Since Engelmann's experiments, it has been shown that in addition to the class of green pigments necessary for photosynthesis, which are known as the chorophylls, there are also yellow, orange, and brown photosynthetic pigments, known as **carotenoids** and **xanthophylls**. Some plants have additional photosynthetic pigments, known as **anthocyanins**, which are stored in large vacuoles. When autumn approaches, certain sugars are converted to anthocyanin pigments that are red under acidic conditions and blue under alkaline conditions, contributing to the fall colors.

PHOTOSYNTHETIC AUTOTROPHS

Unlike organisms known as **heterotrophs**, which require other plants and animals for their livelihood, **autotrophs** are organisms that subsist on the inorganic environment. Autotrophs manufacture organic compounds from molecules that are so small, they don't have to be digested. Because the molecules taken in are small enough and sufficiently soluble to pass through the cell membranes, autotrophic organisms do not need to pre-treat, break down, or digest their nutrients before taking them into their cells.

Photosynthetic autotrophs are among the most important and widespread organisms alive. They have elaborate systems that enable them, with the use of energy from the sun, to raise electrons to an excited state. When the electron is returned to its more normal state, a portion of its energy is transferred to a form where it may be used by the organism.

So green plants and other photosynthetic organisms create high-energy organic material through photosynthesis. As described previously, this involves converting carbon dioxide and water, in the presence of light, into carbohydrate, oxygen, and water. Photosynthesis is the ultimate source of all the energy-rich carbon compounds used by all organisms; it is responsible for the continual supply of atmospheric oxygen, without which all the aerobic organisms, those that use oxygen for all their oxidative processes, would not exist.

The only organisms that photosynthesize are the green plants and algae, some unicellular green flagellates (see Chapters 2 and 18 for more about one-celled organisms with flagella), and two groups of bacteria. Each year these animals, through photosynthesis, release about one-half of all the oxygen that is currently present in the atmosphere. And at the same time, animals, through their respiratory processes, use that oxygen for their metabolism, and replace it with carbon dioxide, which in turn is recycled by the plants.

NUTRIENTS

While animals eat plants, animals, or both, to obtain the nutrients described in Chapter 14, the plants and many microorganisms sometimes obtain their nutrients through inorganic sources, such as the air, water, soil, and the sun. Not really food from a human perspective, the chemicals absorbed by plants and microorganisms are used for metabolic purposes and, therefore, can be categorized as nutrients.

In the same way that the concept of what constitutes food differs, depending on the type of organism being considered, the methods used by organisms to obtain their energy and nutrients also differ. Like all organisms, plants and microorganisms require carbon, oxygen, hydrogen, and nitrogen. In lesser amounts, they also require phosphorus, sulfur, potassium, calcium, magnesium, and iron, as well as trace elements such as molybdenum, boron, copper, and zinc. And certain algae need vanadium and cobalt.

Of the four major elements found in all organisms—carbon, oxygen, hydrogen, and nitrogen—the oxygen and hydrogen are readily obtained from the air or the water. Oxygen often enters plant tissue through the roots and leaves. Inorganic elements, including nitrogen, usually enter higher plants through the roots, and most plants obtain their carbon through the leaves, usually as carbon dioxide. Other essential elements pass through cell membranes by diffusion or active transport (both are defined in Chapter 2), and some cells may ingest particulate matter by pinocytosis and phagocytosis (also defined in Chapter 2).

CHLOROPLASTS

As stated in Chapter 2, plastids are the relatively large organelles in plant cells where nutrient storage and/or photosynthesis occurs. **Chloroplasts**, the plastids containing chlorophyll, are enclosed within an outer envelope composed of two membranes, an outer and inner membrane. Usually quite large and conspicuously green, chloroplasts can be seen under a light microscope.

In the chloroplasts are thin, flat, plate-like photosynthetic membranes called **lamellae**, or **thylakoids**, located in a protein-rich solution called the **stroma**. These photosynthetic membranes are arranged in stacks, called **grana**, throughout the chloroplasts (see Figure 6.2). Within each chloropast, all the photosynthetic membranes are connected, and they surround an interior space that contains hydrogen ions, which are necessary for the synthesis of ATP molecules. Within the thylakoid membranes are chlorophyll and other light-trapping photosynthetic pigments. These pigments are composed of the molecules involved in the electron transport system and the ATP and $NADPH_2$-synthesizing complexes. (See Chapter 5.)

The photosynthetic thylakoid membranes are arranged in a way that creates considerably more surface area in relation to the total enclosed volume, a key ratio that allows the rapid buildup of hydrogen ions by the activities of all the membraneous surface area. The high relative surface area is also important in allowing the photosynthetic pigments to intercept much of the light energy passing through the leaf, or a specific structure containing the chloroplasts.

The enzymes involved in moving carbon dioxide molecules into carbohydrate molecules are located in the protein-rich stroma surrounding the thylakoids. The ribosomes and DNA contained within the chloroplasts are also located within the stroma (see Chapter 2). Generally, eukaryotic green algae cells contain between 1 and 40 chloroplasts. Prokaryotic cyanobacteria (also known as blue-green bacteria, and sometimes still called blue-green algae) contain photosynthetic membranes throughout their interior, rather than within distinct chloroplastic organelles.

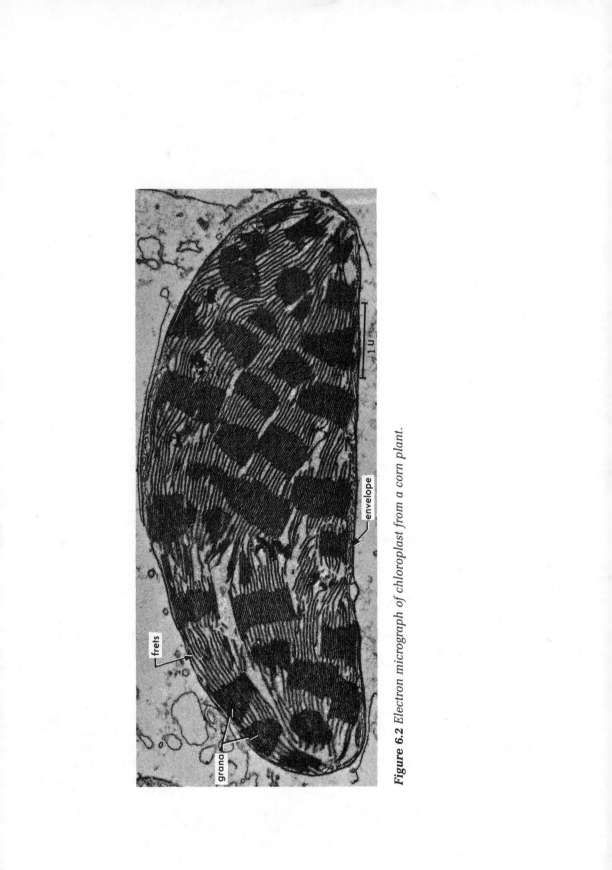

Figure 6.2 Electron micrograph of chloroplast from a corn plant.

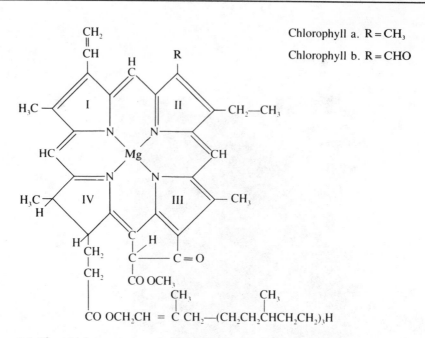

Chlorophyll a. R = CH₃

Chlorophyll b. R = CHO

Figure 6.3 *Chemical structures of chlorophyll* a *and chlorophyll* b.

CHLOROPHYLL AND OTHER PHOTOSYNTHETIC PIGMENTS

There are several different types of chlorophyll molecules, all of which are evolutionarily related. They are referred to as chlorophyll *a*, chlorophyll *b*, and so on. Each is quite similar in structure, having two distinct parts (see Figure 6.3). A long nonpolar end is fat soluble and is anchored within the lipids composing the photosynthetic membranes. Lipids constitute important molecular components of cell membranes. Endoplasmic reticulum, cell membranes, membraneous envelopes surrounding the cell's organelles, and the membranes within the chloroplasts are all largely composed of lipids, and it is within this membrane in the chloroplasts that the long nonpolar tail of the chlorophyll molecules are attached.

The other end of the chlorophyll molecules contains a complex ring structure with a magnesium ion at the center, which is the active site where the light energy is trapped.

In addition to the structure shown in Figure 6.3, there are actually three or four variations with slightly different absorption spectra. The absorption spectrum of a photosynthetic pigment refers to the different wavelengths of electromagnetic radiation—in this case, light, from the sun—that are absorbed by the specific pigment in question. Figure 6.4 illustrates the absorption spectra of chlorophyll *b*.

Because each photosynthetic pigment absorbs light most efficiently within a specific spectrum, plants have several different photosynthetic pigments, increasing the overall efficiency by capturing light energy from a wider range of wavelengths. For example, the carotenoids, another important group of acces-

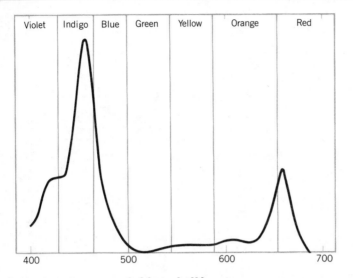

| Violet | Indigo | Blue | Green | Yellow | Orange | Red |

400   500   600   700

Figure 6.4 *Absorption spectrum of chlorophyll* b.

sory pigments found in all green plants, absorb energy that is then passed to the chlorophyll molecules, where it is used in photosynthesis. The carotenoids are long molecules consisting of chains of carbon and hydrogen (hydrocarbon chains) with many double bonds throughout and specific attached side groups, such as methyls, and usually there are ring structures at both ends (see Figure 6.5).

Oxygen molecules are reduced in intense light; this means electrons are usually added to the resulting **free radicals**, which in this case are oxygen molecules with an odd, or unpaired, electron. These radicals are extremely reactive because of their tendency to gain or lose electrons and, therefore, they can react with and destroy other molecules. However, because the free radicals rapidly bind with the double bonds in the carotenoid molecules, they are prevented from destroying the chlorophyll.

If the light intensity is too strong, or sustained over too long a period, the backup system may be inadequate and the chlorophyll molecules will be destroyed by the free radicals. In many plants, chlorophyll is broken down in autumn before the onset of cold winter weather. At this time, before certain plants lose their leaves, the chlorophyll is digested and the magnesium and nitrogen are transported to the roots, where they will be stored until the following spring, when they are sent back up the plant and are used. It is the decomposition of chlorophylls that allows the yellows, oranges, and browns of previously masked carotenoids to become visible. And, as mentioned earlier, the colors from the xanthophylls and anthocyanins may also become obvious. Such color changes sometimes signal certain animals, such as birds, luring them in to eat the ripened fruit and distribute the seeds via their feces, elsewhere.

In addition to the photosynthetic pigments discussed above, red algae and blue-green bacteria contain **phycobilins**, which absorb light energy from

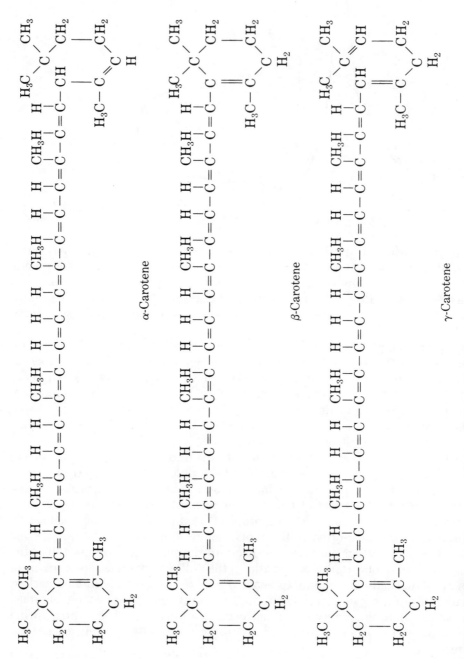

Figure 6.5 Chemical structure of three closely related carotenoids: α-, β-, and γ- carotene.

wavelengths outside the absorption spectra of chlorophyll *a*, and then they transfer the energy to chlorophyll *a* for use in photosynthesis.

LIGHT ABSORPTION

When light energy (photons) is absorbed by the photosynthetic pigments, which are located within the thylakoid membranes, the excess energy in these excited molecules is passed on to chlorophyll *a* molecules. For instance, when light strikes a pair of chlorophyll molecules, the electron held between them absorbs this energy, raising it from its normal stable energy level to a higher energy state. This electron then jumps to another molecule. The chlorophyll molecules are left with a net positive charge, which is then neutralized with an electron that comes from either a nearby water-soluble moleculr or an electron jumping from another photosynthetic pigment molecule, and the process occurs again.

The energized electrons coming off the chlorophyll molecules get passed from one pigment molecule to the next until they reach either of two specialized forms of chlorophyll *a*, called **P680** or **P700**. The P is an abbreviation for pigment; P680 has a maximum absorption peak in the red light with a wavelength of 680 nm. P700 has a slightly longer absorption peak of about 700 nm. (1 nanometer (nm) = 1 millionth of a millimeter (mm).) Both of these specialized chlorophyll molecules have light absorption peaks in the "long" wavelength end of the spectrum, where the energy is considerably less than at the shorter wavelength end of the spectrum. Higher energy light with shorter waves is absorbed by other photosystem pigments and then passed down the energy gradient, eventually being trapped at the low-energy end by either P680 or P700. There are discrete **photosystems**, each of which contains about 200 molecules of chlorophyll *a*, about 50 molecules of carotenoid pigment, and one molecule of either P680 or P700. The former is called **photosystem I** and the latter is **photosystem II**.

PASSING THE ELECTRONS

As the light energy is converted into excited electrons that are passed down the chain of pigment molecules, a series of chemical reactions, known as **redox reactions**, is triggered. Redox is short for **reduction** and **oxidation** reactions; reduction means the addition of an electron (storing energy), and as explained more fully at the beginning of Chapter 5, oxidation means the removal of an electron (releasing energy). Since an electron moving from one molecule to another continually takes energy from one molecule and adds it to another, it follows that as one molecule is reduced, another is oxidized.

This electron transport system ultimately stores light energy two different ways. Electrons are taken from P680 or P700 molecules by strong electron acceptors (referred to either as Z or as FRS—ferredoxin-reducing substance). Z passes electrons to ferredoxin, another iron-containing electron acceptor. And when the electrons reach the outer surface of the thylakoid membrane, the stroma, they are passed to the electron acceptor, a hydrogen-carrying coenzyme **NADP** (nicotinamide adenine dinucleotide phosphate). (NADP is also known as TPN, triphosphopyridine nucleotide.) The NADP retains a pair of energized electrons and, in this state, an NADP pulls two hydrogen protons, H^+, from water to form $NADPH_2$.

Instead of using the notation $NADPH_2$, texts often use $NADPH + H^+$, which indicates that in addition to the NADPH, a hydrogen ion was also added. Either form designates reduced NADP (when the hydrogens are added on, so too are electrons, maintaining a neutral charge).

The source of electrons constantly moving through the system is the accumulation of hydrogen ions in the interior space of the thylakoid membranes, known as the hydrogen ion reservoir. Some of these hydrogen ions are produced by splitting water molecules, and this is where the waste product, oxygen, is produced.

$$2H_2O \rightarrow 4H^+ + 2e^- + O_2$$

The $NADPH_2$ is valuable because of its great reducing potential. When it reduces another molecule by donating its electrons (at the same time it gives up its hydrogens), it releases about 50 kcal of energy per mole.

In photosystem II the photosynthetic pigments trap the energized electrons that get passed to the **electron acceptor Q**, which passes them through a chain of acceptor molecules. While the electrons move along this chain, some energy is released that is used to synthesize ATP from ADP. ATP is a high-energy compound that provides energy for most of the work done by the cell. ATP is a nucleotide, a 5-carbon sugar molecule with a phosphate group and a purine or pyrimadine attached. Nucleotides are the building blocks that make up nucleic acids. Two prominent nucleic acids are DNA (deoxyribonucleic acid) and RNA (ribonucleic acid).

The energy supply required for making the ATP molecules that do the cell's work ultimately comes from the sun, via photosynthesis. Respiration releases the energy from the food molecules manufactured during photosynthesis that store the energy from the sun.

To synthesize ATP, ADP must be joined to an inorganic phosphate group P. In reverse, ATP is converted to ADP plus an inorganic phosphate group and energy. The structural formula of ATP is illustrated in Figure 6.6. The bonds illustrated with wavy lines in the ATP structural formula are referred to as high-energy bonds; they are simply more accessible and thus more apt to break than those bonds signified with either single or double lines.

The adenine and the ribose sugar complex compose the adenosine unit. Attached are three phosphate groups; the last two are connected by high-energy bonds. When broken, the resulting products are ADP and a phosphate group, as well as the released energy. Each may be summarized by the following shorthand.

$$ATP = adenosine-P \sim P \sim P$$

$$ATP \xrightarrow{\text{enzyme}} ADP + P + energy$$

$$ADP + P + energy \xrightarrow{\text{enzyme}} ATP$$

The sequence of events following the movement of electrons through the photosynthetic system may be summarized as follows.

$$H_2O \rightarrow photosystem\ II \rightarrow Q \rightarrow carrier\ chain \rightarrow$$

$$photosystem\ I \rightarrow Z \rightarrow ferredoxin \rightarrow NADPH_2 \rightarrow carbohydrate$$

NONCYCLIC PHOTOPHOSPHORYLATION

Noncyclic photophosphorylation is the process of synthesizing $NADPH_2$ in which the high-energy chlorophyll molecule initially donates electrons and then accepts them when in a low-energy state. This photophosphorylation is called noncyclic because the same electrons are not continually passed around the system; rather an outside source is required. The following equation summarizes this chemical reaction:

$$2ADP + 2P + 2NADP + 2H_2O + light\ energy \rightarrow 2ATP + 2NADPH_2 + O_2$$

CYCLIC PHOTOPHOSPHORYLATION

In the **photosynthetic phosphorylation** process, the light energy–driven addition of phosphate groups described above, part of the overall set of reactions is known as **cyclic photophosphorylation**. Chlorophyll acts as both an electron donor and acceptor. The electrons are passed from molecule to molecule in a chain of reactions. Each step of the way, the electron loses some of its energy. Finally, when the electrons return to the chlorophyll molecules, they have lost their extra energy.

| ATP | Glucose | Glucose phosphate | ADP |

Figure 6.6 *The transfer of the ATP molecule's terminal high-energy phosphate to a glucose molecule. This energizes the glucose and converts it to glucose phosphate, which is important in additional chemical reactions.*

The light strikes the chlorophyll molecules in photosystem I, causing the electrons to become excited. In turn, the electrons then pass along to the electron acceptor molecule Z, which passes them to other acceptor molecules. Each step follows a downward energy gradient. The released energy fuels the synthesis of ATP from ADP and inorganic phosphate. Eventually the electrons are returned to the chlorophyll molecules where they started. The term *cyclic* photophosphorylation refers to this cycling of electrons.

CARBOHYDRATE SYNTHESIS

That light energy is captured and used to make high-energy molecules of ATP and $NADPH_2$ has been explained above. These energy-rich compounds help synthesize carbohydrates, which are the major endproducts of photosynthesis. The photosynthetic processes in which ATP and $NADPH_2$ are synthesized occur in the presence of light.

Then, with these energy-rich compounds, the synthesis of carbohydrates is carried out in either light or darkness. Cells furnished with the proper biochemical apparatus can synthesize carbohydrates at any time, just as long as they are healthy cells, functioning within the proper temperature range, and are furnished with CO_2, ATP, and $NADPH_2$.

Both ATP and $NADPH_2$ fuel the synthesis of carbohydrates. CO_2 is pushed up an energy gradient and converted into a series of intermediate compounds until a 3-carbon sugar, **PGAL (phosphoglyceraldehyde)**, is formed. See Figure 6.7, which illustrates how these processes work together to synthesize this carbohydrate. Some of the PGAL is used to make **ribulose**, the 5-carbon sugar that combines with CO_2, making a 6-carbon sugar that is promptly broken into two 3-carbon sugars, or **PGA (phosphoglyceric acid)**, which is converted into energy-rich PGAL.

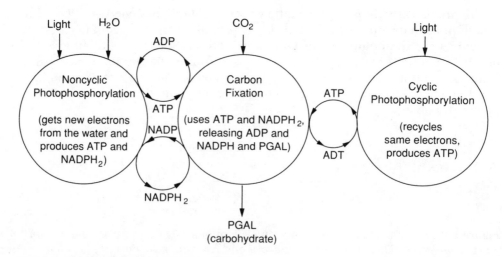

Figure 6.7 *Carbohydrate synthesis.*

Much of the PGAL goes back into more ribulose to combine with the CO_2 molecules. Some PGAL goes straight into the metabolism of the cell, and some, through a series of steps, is combined and rearranged to form the 6-carbon sugar, **glucose**, which most people recognize as the final product of photosynthesis.

The glucose is then available to be broken down for its energy, which is released and used in the cell's metabolic processes. Other glucose molecules can be used to synthesize additional types of molecules such as fats, or they can be strung together to make more complex carbohydrates, such as sucrose, starch, or cellulose. Sucrose, a water-soluble disaccharide, is the sugar transported in solution through the vascular tissue of plants. Starch is the insoluble carbohydrate that is commonly stored in parts of plants such as the roots.

LEAVES

Most of a plant's surface area contributing to the photosynthetic process occurs in the leaves. Hundreds of millions of years of natural selection have shaped leaves into efficient structures that maximize their exposure to light, control their gas exchange, minimize water loss, and help move water, minerals, and carbohydrates up from the roots and to other parts of the plant.

About 10 percent of all the photosynthesis that occurs is the product of higher plants, those that we regularly see around us, and the other 90 percent is the product of algae, most of which occur in the ocean. However, it is the leaves of terrestrial plants that are presented in transverse section in most biology courses and texts, and Figure 6.8 carries on the tradition. Other related structures are described in Chapter 19.

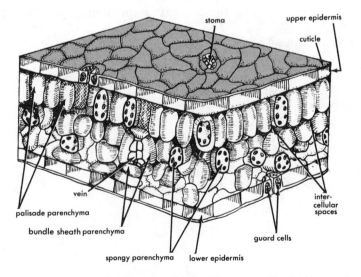

Figure 6.8 *Three-dimensional diagram of a leaf section, illustrating external and internal structures.*

Key Terms

anthocyanins
ATP
autotroph
carbon fixation
carbohydrate synthesis
carotenoids
chlorophylls
chloroplast
cyclic photophosphorylation
electron acceptor Q
free radicals
glucose
grana
heterotroph
Hooke
Ingen-Housz
lamellae
NADP
noncyclic photophosphorylation
oxidation

P680
P700
phosphoglyceraldehyde (PGAL)
phosphoglyceric acid (PGA)
photosynthetic autotroph
photosynthetic pigments
photosynthesis
photosystems
phycobilins
Priestly
redox
reduction
ribulose
Schleiden
Schwann
Senebier
stroma
thylakoid
xanthophylls

Chapter 6 Self-Test

QUESTIONS TO THINK ABOUT

1. Give a brief history of how the basic fundamentals of photosynthesis were first discovered.
2. What are the different photosynthetic pigments and why are there different ones?
3. Describe the differences between autotrophs and heterotrophs.
4. List some of the most important nutrients to a plant, and tell where they are most likely to come from.
5. How does chlorophyll work?
6. Where is chlorophyll usually located, and why?
7. What is noncyclic photophosphorylation?
8. What is cyclic photophosphorylation?
9. How do plants synthesize carbohydrates?

MULTIPLE-CHOICE QUESTIONS

Introduction and Chloroplasts

1. The process fueled by the sun, where carbon dioxide and water are transformed into carbohydrates is known as _____ .
   a. respiration
   b. oxidation
   c. reduction
   d. carbosynthesis
   e. photosynthesis

2. The following investigator demonstrated in 1772 that a plant alone in an airtight jar will die, and an animal alone in an airtight jar will die. However, when a plant and an animal are both placed together in an airtight jar, both survive.
   a. Hooke
   b. Schwann
   c. Schleiden
   d. Priestly
   e. Senebier

3. In 1779, the following researcher showed that sunlight was necessary for plants to produce oxygen:
   a. Hooke
   b. Schwann
   c. Priestly
   d. Senebier
   e. Ingen-Housz

4. In 1782, the part-time scientist _____ showed that plants use carbon dioxide when they produce oxygen.
   a. Schleiden
   b. Priestly
   c. Senebier
   d. Ingen-Housz
   e. Schwann

5. In 1804, _____ found that water is necessary for the photosynthetic production of organic materials.
   a. Priestly
   b. Senebier
   c. Ingen-Housz
   d. Engelmann
   e. de Saussure

6. In 1883, the following researcher conducted an experiment that provided circumstantial evidence indicating chlorophyll, the green pigment
in plants, might be important to photosynthesis:

    a. Priestly
    b. Senebier
    c. Ingen-Housz
    d. Engelmann
    e. de Saussure

7. The following are photosynthetic pigments:

    a. chlorophyll a and b
    b. xanthophylls
    c. anthocyanins
    d. carotenoids
    e. all of the above

8. Based on their mode of nutrition, the following category of organisms
subsists on the inorganic environment, taking in small molecules that do
not have to be digested, from which they manufacture organic compounds:

    a. inorganotrophs
    b. organotrophs
    c. homotrophs
    d. heterotrophs
    e. autotrophs

9. The relatively large organelles in plant cells where nutrient storage and/or
photosynthesis occur are known as _____ .

    a. thylakoids
    b. lamellae
    c. chlorophyll
    d. plastids
    e. none of the above

10. The plastids containing chlorophyll are called _____ .

    a. thylakoids
    b. lamellae
    c. chloroplasts
    d. all of the above
    e. none of the above

11. The thin, flattened sacs inside the chloroplasts are called _____
_____ .

    a. thylakoids
    b. lamellae
    c. stroma
    d. a and b
    e. a and c

12. The stack-like groupings of the photosynthetic membranes located inside the chloroplasts are known as ————————— .
    a. grana
    b. stroma
    c. plastids
    d. chloroplasts
    e. all of the above

13. Surrounding the thylakoids is a protein-rich solution, the ———— ————— , which contains enzymes involved in moving carbon dioxide molecules into carbohydrate molecules.
    a. grana
    b. stroma
    c. plastids
    d. chloroplasts
    e. none of the above

Chlorophyll and Light Absorption

14. The green pigments necessary for photosynthesis are known as ————— ———— .
    a. xanthophylls
    b. carotenoids
    c. chlorophylls
    d. anthocyanins
    e. all of the above

15. There are several different types of chlorophyll molecules; each is quite ————————— .
    a. similar
    b. different
    c. red
    d. blue
    e. orange

16. Chlorophyll molecules have two distinct parts. There is a long nonpolar end that is soluble and is anchored within the lipids composing the photosynthetic membranes. The other end of the chlorophyll molecules contains a complex ring structure with a ————————— ion at the center, which is the active site where the light energy is trapped.
    a. iron
    b. cadmium
    c. calcium
    d. manganese
    e. magnesium

17. _____ refers to the different wavelengths of electromagnetic radiation that are absorbed by the specific pigment in question.

 a. x-rays
 b. gamma rays
 c. ionizing radiation
 d. prismatic spectrum
 e. absorption spectra

18. The _____ are an important group of accessory pigments found in all green plants. There is evidence that they absorb energy which is then passed to the chlorophyll molecules that are used in photosynthesis.

 a. free radicals
 b. phycobilins
 c. nucleotides
 d. carotenoids
 e. adenoids

19. Chlorophyll may be broken down in the autumn before the onset of winter. At this time some plants digest their chlorophyll to save the _____ and _____ atoms which are then transported and stored in their roots.

 a. calcium, iron
 b. hydrogen, oxygen
 c. nitrogen, magnesium
 d. chlorine, potassium
 e. carbon, iodine

20. In addition to the photosynthetic pigments found in most advanced plants, red algae and blue-green bacteria contain _____ which absorb light energy from wavelengths outside the absorption spectra of chlorophyll a, and then transfer this energy to chlorophyll a to be used in photosynthesis.

 a. carotenoids
 b. xanthophylls
 c. anthocyanins
 d. phycobilins
 e. none of the above

Light Absorptions and the Passing of Electrons

21. When light is absorbed by the photosynthetic pigments which are located within the _____ , the energy in these excited molecules is passed on to chlorophyll a molecules.

    a. nucleus
    b. endoplasmic reticulum
    c. thylakoid membranes
    d. adenoids
    e. none of the above

22. The energy supply required for making the ATP molecules, that do the cell's work, either comes from photosynthesis, which captures and stores the sun's energy in food molecules, or it comes from _____ , the process that breaks down food molecules, releasing their energy.

    a. respiration
    b. transpiration
    c. ingestion
    d. refaction
    e. inspiration

23. Cyclic photophosphorylation and noncyclic photophosphorylation may occur in _____ .

    a. the dark
    b. the light
    c. roots
    d. all of the above
    e. none of the above

24. ATP and NADPH$_2$ are synthesized in the _____ .

    a. dark
    b. light
    c. roots
    d. all of the above
    e. none of the above

25. With ATP and NADPH$_2$, the synthesis of carbohydrates can be carried out. Unlike photophosphorylation, carbohydrate synthesis can occur in the _____ .

    a. dark
    b. light
    c. bark
    d. all of the above
    e. none of the above

ANSWERS

1. e	8. e	15. a	22. a
2. d	9. d	16. e	23. b
3. e	10. c	17. e	24. b
4. c	11. d	18. d	25. a
5. e	12. a	19. c	
6. d	13. b	20. d	
7. e	14. c	21. c	

CHAPTER 7

Homeostasis

FLUID ENVIRONMENT OF THE EARLIEST CELLS

The intracellular conditions of most organisms are remarkably uniform, indicating they have changed little throughout evolutionary history. Because maintenance of a stable internal environment, known as **homeostasis,** is critical to the well-being of the organism, regulatory mechanisms are carefully controlled. In Chapter 8, on the endocrine system, the hormones that help regulate this carefully mediated series of feedback systems are discussed.

All life forms are thought to share a marine origin because the fluids found inside most cells have a salt concentration comparable to that of salt water. The reasoning goes that if cells first evolved in a marine environment, then they probably were adapted to a high salt concentration, both inside and out. As it turns out, the majority of all unicellular eukaryotes, as well as invertebrates and vertebrates inhabiting salt water, are composed of cells whose internal environment closely resembles the concentration of the salt water they are bathed in. Of the hundreds of thousands of organisms that do not live in the oceans, most have high salt concentrations in their cellular fluids, as well as high concentrations of other elements found in salt water.

In contrast to the majority of freshwater environments, the ocean presents a stable environment with regard to salt concentration, pH, and temperature. Most organisms are very sensitive to even slight changes in their external environment and are unable to survive seemingly minor changes in their surrounding medium. Just a slight variation in the blood's potassium concentration, for instance, is enough to stop the heart, and a small increase in magnesium will block all nervous activity.

PLANT CELLS VS. ANIMAL CELLS

Algal cells regulate only their intracellular fluids, whereas the cells of some higher plants and most animal cells regulate the fluids both inside and outside their cell membranes. In most animals, approximately equal amounts of fluid are found both intra- and extracellularly (inside and outside the cells), while most of the fluids found in multicellular algae occur intracellularly. Part of this contrast is attributable to the inherent difference between how algae and animals regulate their fluids. In higher plants (those with vascular tissues, e.g., xylem and phloem), the fluids found inside the vascular tissues are essentially continuous with the water outside the plant. Water continually flows from outside the roots, right through the porous root hairs. Animal tissues, in contrast, are separated from the environment by membranes that help regulate the extracellular fluids.

The fluids bathing the cells inside most animals are usually **isosmotic** with the cells; that is, the fluids inside and outside of the cells have about the same osmotic pressure. A large relative osmotic pressure shift, or even in some cases just a slight one, could cause serious problems leading to death (see Figure 7.1). To cope with such a potential shift, many organisms have specialized physiological mechanisms that help them adjust to changes in osmotic pressure; some will be discussed below.

When the intracellular fluids are considerably more concentrated or hyperosmotic relative to the extracellular fluids, animal cells work constantly to

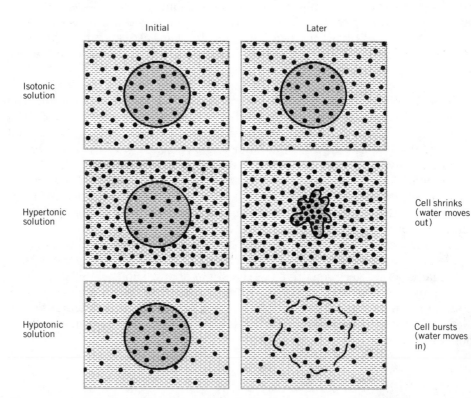

Figure 7.1 *Osmosis is a specialized type of diffusion involving a semipermeable membrane, one that permits the passage of some types of molecules but not others. When a cell, or in the case of most laboratory experiments, a cellophane bag solution with either the same (isotonic), greater (hypertonic), or less (hypotonic) osmotic pressure, the ultimate effects that occur in time are shown in the "later" part of this figure.*

remove inflowing fluids. If the cell cannot remove enough fluid, eventually it will burst due to so much internal pressure.

Generally, plant cells lack the same mechanisms for such rapid removal of material across their membranes; however, most plant cells have a hard cell wall that can withstand the internal pressure that would rip an animal cell apart. So once a plant cell expands to its maximum limit, it remains intact and resists further fluid intake.

When the extracellular fluids surrounding plants become hyperosmotic relative to the fluids inside the cells, the cell walls keep their shape while the cells lose water. Instead, the cell membrane shrinks, pulling away from the cell wall. Such a cell is plasmolyzed, and this process is called **plasmolysis.**

Animals must excrete nitrogenous wastes that result from the breakdown of protein. Terrestrial plants, however, have a greater capacity to convert the toxic wastes that most animals have to excrete into forms that are useful. This is why only animals excrete such nitrogenous wastes as urine and related substances (which are discussed below). If land plants did not have this capacity, they might poison the soils they grow on.

INVERTEBRATE EXCRETION

Contractile Vacuoles

Many unicellular organisms simply pass their excess water, salts, and nitrogenous wastes across their cell membranes into the surrounding medium. However, some protozoans have a specialized excretory organelle, the **contractile vacuole,** that fills with liquid and then moves to the cell membrane, where the contents are released from the cell (or excreted). Rather than being involved in excretion of nitrogenous wastes, contractile vacuoles seem to be primarily involved in eliminating excess water that enters the cell.

Flame Cells

Flatworms such as planaria, flukes, and tapeworms usually have two or more networks of thin tubes running the entire length of their bodies. Together, these constitute the closest thing to the most primitive, multicellular excretory systems that occur in the lower invertebrates. Some of these tubes open to the body's surface through minute pores. Inside these tubes are tufts of cilia that beat continuously in a manner that, when viewed under the microscope, looks like a flickering flame, which is why they are called **flame cells** (see Figure 7.2). Some metabolic wastes are excreted through this flame cell system, but for the most part, as with the contractile vacuoles of protozoans, flame cells seem to be primarily involved in regulation of water balance. The majority of metabolic wastes produced by flatworms are excreted straight from their tissues into their gastrovascular cavity and then out their mouth.

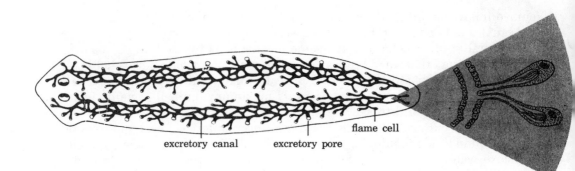

Figure 7.2 *Excretory system of planaria, consisting of flame cells, excretory canals, and excretory pores.*

Nephridia

Unlike flatworms, which lack a circulatory system, annelids (segmented worms) have a closed circulatory system and, therefore, it has been possible for an excretory system to evolve that removes material from the blood, rather than straight from the tissues. Annelids are compartmentalized into many body segments, and each segment is usually equipped with a pair of specialized excretory organs, **nephridia,** which are closely associated with both blood capillaries and pores that open to the outside of the body. The nephridia excrete wastes that they filter from the blood. In addition to these, annelids also pass unused waste products not absorbed in their alimentary canal, through their anus.

Malpighian Tubules

Insects are much more closely related to annelids than to vertebrates, but their excretory systems are completely different from either group, primarily because they have an open circulatory system. It appears that the similarities between nephridia and kidneys are not due to annelids and vertebrates being closely related; they are not. Rather, their excretory systems' similarities appear to stem from both groups having closed circulatory systems. Insects probably evolved from organisms with closed circulatory systems. The closed circulatory system was lost as an open circulatory system predominated, requiring a completely different excretory system.

Their excretory organs are known as **Malpighian tubules.** As waste products accumulate in the hemolymph (or blood) inside the open sinuses, the blood is continuously filtered through a system of tubules that removes the nitrogenous wastes that have been converted into **uric acid crystals.** The tubules doing the filtering connect the open sinuses with the hind gut, where they deposit the waste materials. In the hind gut there is even more resorption of body fluids, so some insects pass out feces that amount to little more than a dry powder (or little, dry pellets).

Kidneys

Unlike annelids, vertebrates don't have excretory organs in each body segment; in fact, vertebrates aren't segmented, and they have just one pair of **kidneys.** Kidneys are large areas where the smaller functional units, known as **nephrons,** are congregated. There are approximately one million nephrons in a human kidney; each constantly filters blood. All the filtrate eventually is excreted in the **urine.**

The blood passes through the **renal artery** (the artery that goes into the kidney). The artery branches into smaller arterioles and then into a small spherical ball of capillaries called the **glomerulus,** which is surrounded by the part of the filtering system called **Bowman's capsule** (or renal capsule, or nephric capsule). When the blood is in the glomerulus, it is under pressure and, since the capillary walls are so thin, much of the plasma passes from the capillary into the capsule. This fluid then passes into the long tube (the nephron—see Figure 7.3). Each section of the nephron is named according to function: filtering out or absorbing specific materials. The first part of the nephron closest to Bowman's capsule is the **proximal convoluted tubule,** the loop in the nephron is called the **loop of Henle,** and the top of the loop on the other side is called the **distal convoluted tubule**. Then the filtered fluid passes into the **collecting duct,** into **larger collecting ducts,** and then into the **ureter**, which takes the urine from each kidney to the **bladder,** where the urine is stored until it is passed through the **urethra** and excreted from the body. All the loops of Henle, each with the same architecture, function as simple mechanical filters; the force that operates them is **hydrostatic pressure.** In addition to removing dilute nitrogenous wastes and excess water, kidneys are also instrumental in regulating the relative concentrations of specific **inorganic ions** in the blood plasma (such as sodium, potassium, and chloride).

This type of kidney probably arose in marine vertebrates, and then helped the first terrestrial vertebrates filter their blood while resorbing much of the water. It has been found that vertebrates inhabiting extremely arid environments have highly specialized kidneys with very long loops of Henle that have a remarkably great capacity to resorb much of the water back into the bloodstream while filtering out just the absolutely necessary materials. These animals have amazingly concentrated urine. Nevertheless, even though mammalian kidneys are well adapted to terrestrial life, they pass so much water that it is usually other vertebrates, such as reptiles, that are far more numerous in arid environments. The reasons are discussed below.

EXCRETION, SALT AND WATER BALANCE

The process of ridding the body of excess water, salts, and nitrogenous wastes such as **ammonia, urea,** and **uric acid** (see Figure 7.4), as well as other useless or toxic substances, is termed **excretion.** In addition to releasing dangerous metabolites, or otherwise useless materials, excretion helps organisms maintain their salt and water balance.

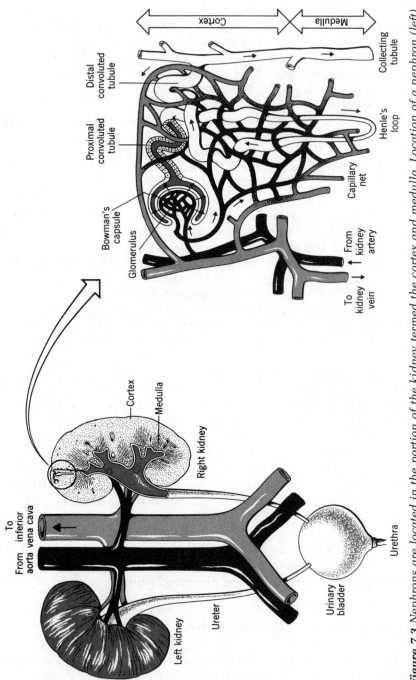

Figure 7.3 *Nephrons are located in the portion of the kidney termed the cortex and medulla. Location of a nephron (left) and its structures (right) are illustrated.*

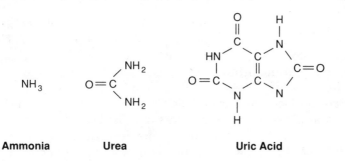

Ammonia   **Urea**   **Uric Acid**

Figure 7.4 *Chemical structures of the nitrogenous compounds: ammonia, urea, and uric acid.*

When carbohydrates and fat supplies are exhausted, the body converts nitrogenous compounds, such as amino acids and proteins, into energy, creating certain nitrogenous by-products. In addition, the body is constantly breaking down and rearranging the components of nitrogenous compounds when manufacturing specific proteins. One of the steps in such conversions involves removing an amino group ($-NH_2$), in a process called **deamination.** During deamination, the amino group is converted into ammonia (NH_3), a toxic waste product that must quickly be converted into a less toxic form.

Only when plenty of fresh water is available, can nitrogenous wastes be sufficiently diluted that they won't poison the animal. Then these wastes can be excreted as ammonia. Many small marine invertebrates pass their wastes across membranes into the surrounding sea water. This is accomplished in some animals without a specialized excretory system. **Phagocytosis,** for example, moves solid wastes across the body wall to the outside, or to the digestive tract, through which they pass and then are excreted. However, such an unspecialized system works in reverse when these organisms are placed in brackish or fresh water. Hypoosmotic media would deplete the organisms' internal fluids, dehydrating them to the point of death, if there weren't adequate compensatory processes.

When vertebrates moved from the water and began colonizing terrestrial habitats, there was no longer any osmotic advantage to maintaining high salt concentrations in their body fluids. Instead, most animals that are adapted to living in fresh water and on land have considerably lower salt concentrations in their internal fluids than their marine ancestors. However, all these animals still have a rather salty internal environment compared to pure, fresh water. When they are placed in fresh water, their internal environment is hyperosmotic in relation to their surrounding medium (meaning their internal environment is saltier than the surrounding fresh water), so there is still a tendency for water to move into them or for salt to move out.

There are many other adaptations that help animals persist in otherwise osmotically inhospitable environments. Some animals, such as certain clams, close up. Others use evasive measures by swimming elsewhere. Many organisms have evolved methods by which they can regulate their internal osmotic concentrations; these animals are said to have the ability to **osmoregulate.**

Some animals, such as most freshwater fish, constantly take in water, because their internal environment is saltier than the water they live in. These animals must constantly excrete the excess water, usually by maintaining a continual flow of extremely dilute urine. Much excess water is also removed through the gills. Salt, lost in the process, is replaced through the food that is eaten, and some salts are replaced by specialized salt-absorbing cells in the gills.

Oddly, it appears that the ancestors of most bony fish once lived in fresh water, and some then became marine. Some descendants of these bony marine fish have retained their relatively dilute bodies and are hypoosmotic relative to marine water. They constantly lose water and must counter dehydration and excessive salt uptake. Their relatively impermeable skin is usually covered with scales, which helps, but since there is so much area for water loss and salt intake through the surface area of the gills, it is a combination of such impermeable skin, with specialized salt-excreting cells in the gills, and the constant drinking of water that enables them to osmoregulate. Much of the nitrogenous waste is passed through the gills, but urine may also be produced by the kidneys.

Sharks also appear to have a freshwater ancestry, and yet they solved their osmotic problems differently. They combine ammonia with carbon dioxide, converting it to urea. But instead of excreting the urea, which in high concentrations is toxic to most other animals, they retain it in their blood, without any adverse effects. It is the urea that enables them to maintain a high osmotic concentration, thereby avoiding dehydration. Whatever salts sharks need to eliminate are excreted through specialized cells in the rectum. It may sound odd that a fish could become dehydrated, but without finely tuned osmoregulation, depending on the species and the environment, fish would easily become either dehydrated or bloated with too much water.

Freshwater animals remove salts by moving them from their tissues to the bloodstream, and then out of the body. Highly specialized salt excretion centers, such as in the gills of many fish, or in the noses or tear ducts of other animals, such as some lizards and birds, help.

OTHER OSMOREGULATORY PROBLEMS

When animals moved to land, some of the osmoregulatory problems encountered were those related to desiccation. Being on land obviously meant the animals were no longer bathed in an aqueous surrounding, so their entire body surface was vulnerable to water loss through evaporation. Some of the first terrestrial forms probably had protective coverings (such as scales) that reduced the potential water loss. Despite protection, however, their respiratory surfaces (lungs, trachea, throat, mouth, and nasal passages) were susceptible to water loss. And additional water was lost through urine and feces. With such a great potential for water loss, few organisms would survive very long without replenishment. So organisms dealt with these problems by drinking fresh water and eating foods containing water, as well as through the oxidative breakdown of nutrients, since water is one of the by-products of cellular respiration.

Some fish and most amphibians and mammals have an adaptation that deals with the ammonia in a manner that conserves water. Like sharks, they combine ammonia with carbon dioxide, converting it to urea. But unlike sharks, the urea is toxic unless in very dilute concentrations. The urea is released into their blood, and then many fish pass the urea directly from their blood out through their gills. Most mammals rely more on their kidneys for the removal of urea.

Unlike animals that lay shelled eggs, the terrestrial animals that lay eggs in water, such as amphibians, are able to use urea as a nitrogenous metabolite because such wastes will readily diffuse into the surrounding water. Mammals, too, have embryos that develop in an aquatic medium that is not enclosed inside a shell, so they too can deal with their nitrogenous wastes by excreting diluted urea.

Because urea still requires some water that otherwise might have been used for other purposes, it is even better for some animals to produce uric acid, which is very insoluble and can be excreted without being diluted, draining the body of water. Apparently, uric acid is necessary for some egg layers, because it can remain inside an egg shell without any toxic effects that would disrupt the embryo's metabolism.

Birds and reptiles convert the ammonia into uric acid, which is released into the blood and removed by the kidneys. Although uric acid is a more complex compound than urea, it does not have to be watered down into a more dilute form to be safe. Therefore, excreting uric acid conserves water, which is especially important for those reptiles that inhabit arid environments. The uric acid crystals excreted by these animals account for their whitish excrement.

Key Terms

ammonia
bladder
Bowman's capsule
collecting duct
contractile vacuole
deamination
distal convoluted tubule
excretion
flame cells
glomerulus
glycogen
homeostasis
hyperosmotic
hypoosmotic
isosmotic
kidneys

liver
loop of Henle
Malpighian tubules
nephridia
nephron
osmoregulation
phagocytosis
plasmolysis
proximal convoluted tubule
renal artery
urea
ureter
urethra
uric acid
urine

Chapter 7 Self-Test

QUESTIONS TO THINK ABOUT

1. Explain what it means for a cell to be isosmotic, hyperosmotic, and hypoosmotic with regard to the surrounding medium.
2. Why is ammonia converted into either urea or uric acid in most organisms?
3. What are the different methods animals have to cope with nitrogenous wastes?
4. Trace the movement of nitrogenous waste removal through a kidney from the renal artery to the ureter.

MULTIPLE-CHOICE QUESTIONS

1. The fluids bathing the internal environment of most animals' cells is usually _____ with the fluid inside the cells.

    a. isosmotic
    b. hyperosmotic
    c. hypoosmotic
    d. hypertonic
    e. hypotonic

2. When plant cells lose water and the cell membrane shrinks, pulling away from the cell wall, such a cell is termed _____ .

    a. turgid
    b. in turgor
    c. plasmolyzed
    d. hypertoned
    e. hypotoned

3. The first step in the conversion of nitrogenous compounds into glucose involves the removal of an amino group, which is termed _____ .

    a. the bends
    b. indigestion
    c. aminization
    d. nitrogen fixing
    e. deamination

4. During deamination, the amino group is converted into _____ .

    a. ammonia
    b. urine
    c. uric acid
    d. urea
    e. amide ions

5. Mammals combine ammonia with carbon dioxide, forming the less toxic nitrogenous compound _____ .

   a. urea
   b. uric acid
   c. amide ions
   d. feces
   e. excrement

6. Birds and reptiles convert the ammonia into _____ .

   a. urea
   b. uric acid
   c. amide ions
   d. water
   e. nitrogen

7. Certain fresh water animals remove salts from their tissues into the bloodstream, from which they are then passed from the body via the

   _____ .

   a. kidneys
   b. gills
   c. nose
   d. tear ducts
   e. any or all of the above

8. Some protozoans have a specialized excretory organelle known as the _____ that fills with liquid that is then carried to the cell membrane where the contents are released.

   a. flame cell
   b. bladder
   c. urethra
   d. contractile vacuole
   e. Malpighian tubule

9. Flatworms and tapeworms usually have a primitive, multicellular excretory system of _____ .

   a. kidneys
   b. Malpighian tubules
   c. flame cells
   d. nephridia
   e. glomeruli

10. Each body segment of an annelid worm is usually equipped with a pair of specialized excretory organs known as _____ .

    a. kidneys
    b. nephridia
    c. Bowman's capsules
    d. glomeruli
    e. flame cells

11. Each functional unit in a kidney is called a ——————— .

    a. nephron
    b. flame cell
    c. bladder
    d. ureter
    e. urethra

12. Insect excretory systems involve organs known as——————— .

    a. flame cells
    b. nephrons
    c. Malpighian tubules
    d. kidneys
    e. ureters

ANSWERS

1. a	4. a	7. e	10. b
2. c	5. a	8. d	11. a
3. e	6. b	9. c	12. c

CHAPTER 8

Hormones

The **endocrine system** consists of a series of tissues, glands, and cells found throughout the body that secrete certain chemicals. These chemicals, called **hormones,** exert specific effects on specific cells and tissues. Some hormones have the potential to significantly affect other parts of an organism's body.

Both plants and animals produce hormones. Those produced by plants emanate primarily from where most of the growth occurs, such as in the buds, seeds, new shoots, and at the root tips. The plant hormone-producing areas have other functions as well. In animals, the sole function of these hormone-producing tissues, or **endocrine glands,** is hormone production. In animals, hormones are distributed through the body via the circulatory system. In vascular plants, hormones are transported by the phloem from the site of synthesis to where they are used.

PLANT HORMONES

Plant hormones have been of considerable interest to researchers because of their many potential practical applications. For instance, when storing fruits, vegetables, and grains, it would be helpful to understand what makes them remain dormant for long periods of time. And because some hormones stimulate the rapid growth of specific plant parts, such as the seeds, or even specific parts of a seed, by manipulating plant hormones it is possible to increase the value of a cereal crop considerably.

AUXINS

The **auxins** represent one of the most widely understood groups of plant hormones. They have been shown to be important in controlling cell elongation in plant stems, especially with regard to varying types of stimuli. For instance, light reduces the auxin supply to the side of a plant it strikes. Since the plant has more auxins on the shaded side, the cells grow faster there, causing the stem to bend toward the light. This bending toward light is known as the **phototropic response.**

Plant **tropisms** usually refer to the turning or bending of a plant part in response to a particular stimulus, such as light, gravity, water, or other nutrients, producing different growth patterns. One such tropism in which auxins are implicated is **geotropism,** which has to do with the direction plant parts grow in response to gravity. A negative geotropic response involves a shoot growing away from the direction in which gravity pulls.

Shoot tips are not only sensitive to light, but they can also detect gravity. When there is an unequal distribution of gravitational pull on all sides, they increase the concentration of auxins on the lower side. This stimulates the cells on the lower side to elongate faster than the cells on the upper side, which gets the plant to grow up again. Roots, unlike shoots, have a positive geotropic response. That is, they turn toward the pull of gravity. Root growth direction is also affected by the concentration gradient of water and specific nutrients.

Auxins are also involved in the inhibition of lateral buds. Those auxins produced in the terminal bud, the bud at the tip of the shoot, move down the shoot and inhibit the development of the nearby buds, while also stimulating the stem to elongate.

It has been demonstrated that the rapid growth of many types of fruit is stimulated by auxins released from the pollen grains that fertilized the ovule (egg), and that, as the seeds develop, they continue to produce more auxins. Auxins are also involved in preventing leaves, flowers, and fruits from falling off the plant. Then, when it is time, hormonal changes such as those triggered by shorter daylength, colder temperatures, or drier conditions, can stimulate the growth of what are known as **abscission layers**, which result in specific plant parts falling off.

In addition to being important in cell elongation and in forming abscission layers, auxins are also involved in cell division. In the early spring, when the auxins move down from the buds, they stimulate the cambium to divide, forming a new layer of xylem. Toward autumn the buds produce less auxin until eventually the production of new buds, leaves, and xylem slows down or stops. Toward the end of winter, renewed auxin production stimulates the resumption of growth.

GIBBERELLINS

The **gibberellins**, another group of plant hormones, probably function in conjunction with, rather than separately from, the auxins. Gibberellins have a dramatic effect on stem elongation, particularly on those plants that are normally known for their "dwarf" varieties. Unlike auxins, gibberellins do not produce the bending movements typical of the phototropic and geotropic responses of shoots and roots. Gibberellins do not inhibit the growth of lateral buds, and they don't prevent leaf abscission. While auxins stimulate the cambium to produce new xylem cells, gibberellins stimulate the cambium to produce new phloem cells. Gibberellins have also been implicated in ending seed dormancy, and they have been shown to induce some biennials, which normally take two years to flower, to flower during their first year of growth. Gibberellins also affect the time when plants flower, depending on both timing and duration of the dark periods of a day (see the section on photoperiodism below).

CYTOKININS AND INHIBITORS

Both auxins and gibberellins have been shown to affect cell division, though they appear to work in conjunction with other substances more directly involved in this process. The **cytokinins** are the compounds known to promote cell division. Other compounds have also been found to be important in inhibiting or blocking cell division activity, thereby maintaining the dormancy of buds, seeds, and shoots.

One such hormone is **abscisic acid,** which has been shown to be involved in inducing abscission. Abscission is caused from the growth of thin-walled cells that result in the falling of a leaf or fruit from the plant. Abscisic acid has also been implicated in plant dormancy, stomatal closure, and growth inhibition.

ETHYLENE

Ethylene is a very volatile compound that has a number of different activities in plants. This plant hormone is involved in fruit ripening. It contributes to leaf abscission and to lateral bud inhibition. In addition, it has been shown that when some trees are attacked by herbivorous insects, ethylene is released, which may trigger nearby trees to manufacture chemicals that will protect them from the insects. Ethylene is also involved in the plant's aging process.

PHOTOPERIODISM

The response by an organism to the duration and timing of light and dark is known as **photoperiodism.** Some plants respond to precise daylight periods. It has been found that rather than day length, it is the length of the night that is critical. But since this was discovered after the following terms were coined, they are still with us.

Short-day plants flower when the day length is below a certain critical value, generally resulting in a plant's blooming either during the spring or the fall. **Long-day plants** bloom when the day length exceeds a specific critical value, which is usually during the summer. And **day-neutral plants** can bloom anytime and may respond to other cues besides the length of the daylight or darkness.

Depending on the particular species, a certain day length causes the leaves to manufacture the hormone **florigen,** which moves to the buds and causes flowering. When leaves are exposed to other specific photoperiods, they destroy florigen, and therefore the plants don't flower. It is not precisely known how gibberellins affect flowering, but it appears to differ with species. In some, the effect of gibberellins is indirect. In others, the gibberellins seem to work in conjunction with florigen to induce flowering. It has been found that plants possess a sensitive pigment, **phytochrome** which responds to the presence or absence of light by measuring the time lapse between the onset of darkness until the next exposure to light. Phytochrome is coupled with florigen synthesis.

ANIMAL HORMONES

Hormones are important to animals as well as plants. Among invertebrates and vertebrates, hormones are involved in the regulation of growth, development, and homeostasis. Specific animal organs produce hormones that travel, usually via the blood, to other organs where they coordinate certain bodily functions. In animals, the release of hormones is usually triggered by nervous stimuli.

There are two basic groups of organs that secrete specific substances into the body. These are the **exocrine glands,** which secrete their products into ducts, which then carry the secretions to the body surface or into the body cavity. Digestive, mucous, sebaceous, and sweat glands are included in this category.

Glands in the other group secrete their products into the general area around the secretory cells, and from this area, the secretions pass into the blood capillaries. These **endocrine glands** are ductless. They include the adrenals, pancreas, pineal, parathyroids, ovaries, testes, thymus, and thyroid. It is the endocrine glands that produce hormones, which are either a protein, an amine, or a steroid (see Figure 8.1).

Regardless of the type of chemical, all hormones, whether protein, amine, or steroid, stimulate cellular changes either in target cells, in a target organ, or in a group of organs. Or the hormone may affect the activities of all the cells in the body. See Table 8.1 for a list of human hormones and their functions.

DIGESTION (GASTROINTESTINAL TRACT)

Passing through the mucosal region of the pyloric sphincter, meat stimulates the release of **gastrin,** a hormone that stimulates the **gastric glands** to secrete **gastric juice,** which starts digesting the meat. Both the stomach and intestine produce gastrin. One is called stomach gastrin and the other, intestinal gastrin.

Fats stimulate the wall of the duodenum to release **enterogastrone**, a hormone that inhibits the secretion of gastric juice. When stimulated by acidic food coming from the stomach, the mucosal cells of the small intestine release the hormone, **secretin**. Secretin stimulates the secretion of **pancreatic juice**. The small intestine, when stimulated by acids and fats, releases the hormone **cholecystokinin (pancreozymin)**, which stimulates the gallbladder to release

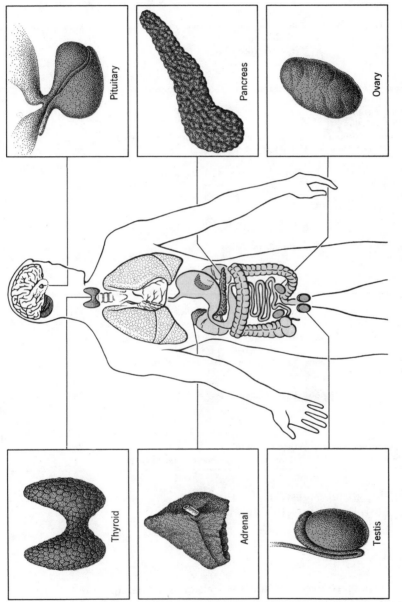

Figure 8.1 *Location and general appearance of important hormone-secreting glands in humans.*

Table 8.1 *Major sources of human hormones and their functions.*

Source	Hormone	Functions	Deficiency	Excess
Thyroid gland	Thyroxine	Stimulate metabolism: regulate general growth and development	Cretinism	Graves' disease
	Calcitonin	Lowers blood calcium		
Parathyroid	Parathormone	Increases blood calcium; decreases blood phosphate	Muscle spasms	Calcium deposits
Pancreas	Insulin	Lowers blood glucose	Diabetes	Hypoglycemia
	Glucagon	Increases blood glucose	Hypoglycemia	
Adrenal medulla	Epinephrine (Adrenalin)	Increases metabolism in emergencies		
	Norepinephrine (Noradrenalin)	As above		
cortex	Glucocorticoids and related hormones	Control carbohydrate, protein, mineral, salt, and water metabolism	Addison's disease	Cushing's syndrome
Pituitary anterior	Thyroid stimulating hormone	Stimulates thyroid gland function		
	Adrenocorticotropic hormone (ACTH)	Stimulates adrenal cortex	Hypoglycemia	Cushing's syndrome
	Growth hormone	Increases body growth	Dwarfism	Gigantism, acromegaly
	Gonadrotropic hormones	Stimulates gonads		
	Prolactin	Milk secretion		
posterior	Vasopressin (ADH)	Water retention by kidneys		
	Oxytocin	Milk production		
Testis	Testosterone (Androgens)	Secondary sex characteristics, sperm production	Sterility	
Ovary	Estrogens	Secondary sex characteristics		
	Progesterone	Prepares uterus for pregnancy		
Hypothalamus	Hypothalamic releasing and inhibiting hormones	Release of hormones from anterior pituitary gland		
Kidney	Renin	Vasoconstriction		Increases blood pressure
	Erythropoietin	Production of red blood cells in bone marrow		
Gut wall	Digestive hormones	Digestion of food		
Thymus gland	Thymosin	Maturation of lymphocyte white blood cells		

bile. Bile aids in fat digestion. It is produced by the liver, stored in the gall-bladder, and released into the duodenum.

HISTAMINE

Damaged tissues release **histamine**, which dilates, or relaxes, the muscles in the walls of blood vessels, thereby making them more permeable to their contents and enabling more white blood cells and antibodies to move into the damaged area to fight infection.

People with certain allergies, such as hay fever, may develop a reaction that causes the nasal mucosa to release histamine. This dilates the nasal blood vessels so that fluids escape from both the blood vessels and the mucosal glands, thus causing a runny nose. This is why such people take **antihistamines.**

PANCREAS

The **pancreas** aids digestion by producing pancreatic digestive enzymes. In addition, the pancreas contains **islet cells,** or **islets of Langerhans,** which produce the hormone **insulin.** This hormone reduces the concentration of glucose in the blood. Too much insulin in one's system, from the rare condition of an overactive pancreas, can produce **insulin shock,** during which the blood sugar level falls so low that a person may become unconscious and die.

More common is the insulin deficiency, known as **diabetes,** that results in the inability of the liver and muscles to control the conversion of glucose into glycogen. Sometimes the liver produces too much glucose from glycogen, depleting all its resources and making the body use its proteins and fats. Diabetes often leads to chronic problems affecting many aspects of one's well-being.

The pancreas also secretes **glucagon.** This has the opposite effect of insulin, causing the amount of glucose in the blood to increase.

ADRENALS AND KIDNEYS

At the anterior (top) end of each kidney is an **adrenal gland** (sometimes called a suprarenal gland), which is composed of the outer **adrenal cortex** and the inner **adrenal medulla.** The cortex produces over 50 different hormones, not all of which are active. All are **steroids,** as are the hormones produced by the gonads (ovaries and testes). The **cortical hormones**, those produced by the adrenal cortex, are grouped according to their function. One group, the **glucocorticoids,** contains hormones that regulate carbohydrate and protein metabolism. Another group, the **mineralocorticoids,** regulate salt and water balance. A third group, the **gonadocorticoids,** consists of certain male and female sex hormones, the estrogens and androgens.

The adrenal medulla secretes **adrenalin** (epinephrine), as well as **noradrenalin (norepinephrine).** Adrenalin decreases insulin secretion, and it also stimulates pulse and blood pressure. In addition, it stimulates the conversion of glycogen into glucose (in the liver), which is then released into the blood. Adrenalin also increases oxygen consumption and the flow of blood to the skeletal muscles (those that move the body), while decreasing the blood flow to the smooth muscles (those involved in digestion). Many of these reactions, which cumulatively are often referred to as the fight-or-flight response, occur when the body is subjected to pain, fear, anger, or other stress. Noradrenalin has similar effects in that it helps to mobilize the body during times of stress.

The kidneys secrete the protein **renin,** which reacts with a blood protein to form the hormone **hypertensin** (also called angiotonin). Hypertensin stimulates the constriction of small blood vessels, increasing blood pressure. This seems to be a response that kidneys use to compensate for reduced blood flow due to blocked arteries. The higher blood pressure can overcome such temporary blockages, allowing the kidneys to filter the necessary amount of blood. Kidneys also secrete **erythropoietin,** a hormone that stimulates red blood cell production.

THYROID

In humans, the thyroid gland, located just below the larynx, around the front and sides of the trachea, produces the hormone **thyroxin,** an amino acid altered with four iodine atoms. Thyroxin's primary function involves the regulation of metabolic activity by increasing the rate at which carbohydrates are burned. Also, it stimulates cells to break down proteins for their energy rather than using them to build new tissues.

Iodine is necessary for proper thyroid function; an insufficient amount in the diet produces a condition known as **hypothyroidism,** resulting in a decrease in energy. Children with this condition can develop improperly. Hypothyroidism can be treated with more iodine in the diet, or with thyroxin.

Hyperthyroidism is due to a thyroid that produces too much hormone, resulting in an increased metabolic rate, higher than normal body temperature, high blood pressure, and weight loss. Elevated thyroid activity can be inhibited with the prescribed treatment of recently discovered drugs.

The thyroid gland releases two other hormones. **Triiodothyronine** is similar to thyroxin, except that it is much stronger. **Thyrocalcitonin** (sometimes called calcitonin) is quite different from the previous two hormones, both structurally and functionally; it helps control the blood calcium level.

PARATHYROIDS

On the thyroid's surface are four small pea-like organs known as the **parathyroids,** which are functionally distinct from the thyroid. They produce the hormone **parathormone,** which regulates the calcium-phosphate balance between the blood and other tissues. A calcium deficiency caused by **hypoparathyroidism** results in nervous twitches, spasms, and convulsions. **Hyperparathyroidism** leads to the demineralization of bone tissue, rendering the bones highly susceptible to fracture.

THYMUS

Located in the upper chest and lower neck, the **thymus gland** is composed of tightly packed lymphocytes. These white blood cells are held in place by fibrous tissue. The thymus is most active from infancy to puberty, after which it atrophies, only to enlarge again in old age. The gland produces **thymosin,** a hormone that stimulates plasma cells in the spleen, lymph nodes, and other lymphoid tissues to function immunologically. Two of the main types of lymphocytes, the B cells and T cells, are produced in the bone marrow and then migrate to lymphoid tissues. Those that end up in the thymus gland become thymus-dependent lymphocytes, or T cells. It may be the thymosin that affects these cells, enabling them to destroy antigens (foreign microbes and substances).

PITUITARY

In the brain is the small gland known as the **pituitary,** also called the **hypophysis,** which consists of two lobes: the **anterior lobe** and the **posterior lobe.** Both are attached via a stalk, the **infundibulum,** to the **hypothalamus,** which is located just above the pituitary. (See Figure 8.2.)

The anterior pituitary, also called the **adenohypophysis,** produces many hormones. Most control the activities of other endocrine glands (see Figure 8.3). Of these hormones, **prolactin** stimulates female mammary glands to produce milk. **Growth hormone** (somatotrophic hormone, STH) is important in regulating growth. **Melanocyte-stimulating hormone** (MSH) triggers pigment molecule dispersion in the pigment-containing cells, often called melanophores in some lower vertebrates such as fish, frogs, and lizards. MSH increases skin pigmentation by stimulating the dispersion of melanin granules in mammal melanocytes; mammals, however, lack melanophores.

Thyrotropic hormone stimulates the thyroid. **Adrenocorticotropic hormone** (ACTH) stimulates the adrenal cortex. **Follicle-stimulating hormone** (FSH) and **luteinizing hormone** (LH) act on the gonads. In females, FSH initiates the development of an ovum each month. In males, FSH stimulates the testes to produce more sperm. In females, LH stimulates the ovary to release the developed ovum. It also stimulates the **corpus luteum** in the ovary to secrete **progesterone,** which prepares the uterus for receiving the embryo.

The posterior lobe, or **neurohypophysis,** contains neuron fibers that connect with the **hypothalamus.** The cell bodies of these neurons produce two

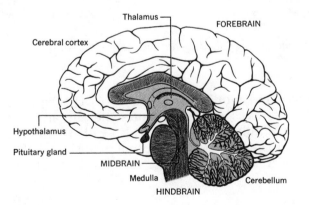

Figure 8.2 Location of the pituitary gland and hypothalamus, as illustrated in this cross-section of a human brain.

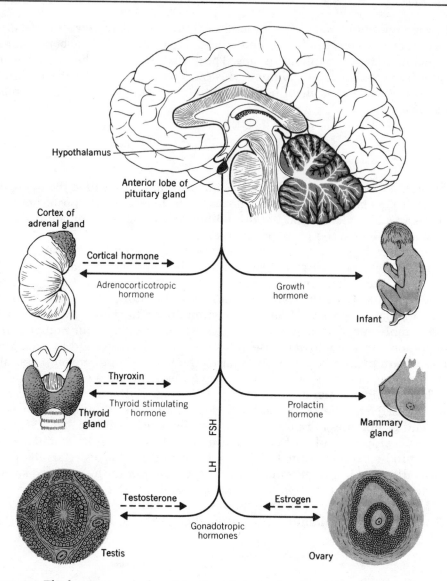

Figure 8.3 *The hormones produced by the pituitary gland and the parts of the body where each has its effect.*

hormones: **oxytocin** and **antidiuretic hormone** (ADH). Oxytocin stimulates the contraction of the uterus as well as the contractile cells around the ducts in the mammary glands. Antidiuretic hormone, also called **vasopressin,** stimulates the kidneys to absorb more water and return it to the blood, thereby decreasing the urine volume. Alcohol inhibits the secretion of ADH, increasing urine output.

HYPOTHALAMUS

The hypothalamus receives nervous impulses from the body's sense organs and then responds by secreting **releasing factors** into the blood. The releasing factors, which are hormone-like substances, stimulate the anterior pituitary to secrete specific hormones.

The onset of puberty, the sequence of events that transforms a child into a young adult, begins when the hypothalamus secretes follicle-stimulating hormone–releasing factor, which signals the anterior pituitary to secrete follicle-stimulating hormone and **interstitial cell-stimulating hormone** (ICSH), both of which are hormones that stimulate the **gonads** (testes and ovaries). Such hormones are referred to as gonadotropic hormones. (See Figure 8.3.)

In the female, the menstrual cycle and ovarian cycle are controlled by the **follicle-stimulating hormone–releasing factor** (FSHRF) and the **luteinizing hormone–releasing factor** (LHRF). Follicle-stimulating-hormone releasing factor stimulates the anterior pituitary to release FSH, which stimulates follicular development and **estrogen** secretion by the follicles.

Luteinizing-hormone releasing factor stimulates the anterior pituitary to release LH, which stimulates the development of ovarian follicles, leading to ovulation. In addition, it stimulates the production of the female growth and maturation hormones, estrogens and progesterone.

OVARIES

In the ovaries are jackets of cells surrounding the potential egg cells. These are known as follicles. They release estrogens, which are involved in the development and maintenance of the uterus and breasts. Estrogens are also involved in fat distribution and deposition, voice pitch, broadening of the pelvis, and hair growth.

The **corpus luteum**, which is formed from the follicle after ovulation, secretes both estrogens and **progesterone**, which prepare the **endometrium**, the internal layer of the uterus, for implantation of an embryo, and it helps prepare the breasts for milk production.

On average, from the first period, or **menses (menarche),** to **menopause**, the termination of menstrual cycles, the **menstrual cycle** lasts 28 days. The menstrual cycle can be divided into the menstrual phase (menses), the pre-ovulatory phase, ovulation, and the post-ovulatory phase. The main events that occur in the ovary and uterus with regard to the above hormones are illustrated in Figure 4.10 on page 77.

During birth, estrogen stimulates the contractions of the uterine muscles, and progesterone inhibits these muscular contractions. Oxytocin is also thought to be involved in uterine contractions. Another hormone, **relaxin,** which is secreted by the ovaries and placenta during pregnancy, helps loosen some of the connections between the pelvic bones, making them more flexible to enlarge the birth canal when the baby is being born. This occurs to a lesser extent in humans than in many other mammal species.

TESTES

Follicle-stimulating hormone stimulates the seminiferous tubules to begin spermatogenesis. Interstitial cell-stimulating hormone assists the seminiferous tubules to develop mature sperm and also stimulates the interstitial cells in the testes to secrete **testosterone.**

Just prior to birth, testosterone stimulates the descent of the testes into the scrotum. Testosterone also controls development, growth, and maintenance of the male sex organs. At puberty, testosterone stimulates the secondary male sex characteristics such as the development of more muscle, more body hair, and deepening of the voice. The **anabolic steriods** used by many athletes to increase muscle mass are artificially manufactured forms of testosterone. One of the possible side-effects may be male sterility.

Key Terms

abscisic acid
abscission layers
adenohypophysis
adrenal cortex
adrenal gland
adrenalin
adrenal medulla
adrenocorticotropic hormone
anabolic steroids
anterior lobe
antidiuretic hormone
antihistamines
auxins
bile
cholecystokinin
corpus luteum
cortical hormones
cytokinins
day-neutral plants
diabetes
endocrine glands
endocrine system
endometrium
enterogastrone
erythropoietin
estrogen
ethylene
exocrine glands
florigen
follicle-stimulating hormone
follicle-stimulating hormone–
  releasing factor

gastric glands
gastric juice
gastrin
geotropism
gibberellins
glucagon
glucocorticoids
gonadocorticoids
gonadotropins
gonads
growth hormone
histamine
hormone
hyperparathyroidism
hypertensin
hyperthyroidism
hypoparathyroidism
hypophysis
hypothalamus
hypothyroidism
infundibulum
inhibitors
insulin
insulin shock
interstitial cell-stimulating hormone
islet cells
islets of Langerhans
long-day plants
luteinizing hormone
luteinizing hormone–releasing factor
melanocyte-stimulating hormone
menarche

menopause
menses
menstrual cycle
mineralocorticoids
neurohypophysis
noradrenalin
ovaries
oxytocin
pancreas
pancreatic juice
pancreozymin
parathormone
parathyroids
photoperiodism
phototropic response
phytochrome
pituitary gland
posterior lobe
progesterone

prolactin
relaxin
releasing factors
renin
secretin
short-day plants
steroids
testes
testosterone
thymosin
thymus gland
thyrocalcitonin
thyroid gland
thyrotropic hormone
thyroxin
triiodothyronine
tropism
vasopressin

Chapter 8 Self-Test

QUESTIONS TO THINK ABOUT

1. What are hormones, and how are they delivered from where they are produced to their target area?
2. What is the difference between endocrine and exocrine glands?
3. What are two groups of plant hormones, and how do they affect plants?
4. What is a plant's photoperiod?
5. Describe how hormones affect diabetes.
6. Compare the similarities and differences of male and female sex hormones.

MULTIPLE-CHOICE QUESTIONS

Introduction, Plant Hormones, Auxins, Gibberellins, Cytokinins

1. Chemicals that exert specific effects on target tissues are called _____ .

    a. hormones
    b. auxins
    c. gibberellins
    d. cytokinins
    e. all of the above

2. The following group of plant hormones, ——————, has been shown to be important in controlling cell elongation in plant stems, as well as in roots, and is also involved in the phototropic and geotropic responses.

a. cytokinins
b. gibberellins
c. auxins
d. ethylenes
e. all of the above

3. A plant —————— is the term usually used to describe the turning or bending of a plant part in response to a particular stimulus, usually due to different growth patterns.

a. tropism
b. gibberellin
c. auxin
d. all of the above
e. none of the above

4. The following group of plant hormones, ——————, has a dramatic effect on stem elongation, particularly on plants that are normally "dwarf" varieties.

a. gibberellins
b. auxins
c. cytokinins
d. all of the above
e. none of the above

5. While both the auxins and gibberellins have been shown to be involved in stimulating cell division, apparently they work in conjunction with other substances more directly involved in this process. The group of compounds that promote cell division are called ——————.

a. inhibitors
b. ethylenes
c. phytochromes
d. cytokinins
e. none of the above

Animal Hormones, Introduction, Digestion (Gastrointestinal Tract), Histamine

6. The following are examples of endocrine glands:

a. adrenals, pancreas, pineal, mucous
b. parathyroids, ovaries, testes, sweat
c. thymus, thyroid, adrenals, testes, mucous
d. sweat, pancreas, pineal, ovaries, thymus
e. parathyroids, testes, thymus, thyroid, adrenals

7. Endocrine glands produce hormones which are _____ .

   a. proteins
   b. steroids
   c. amines
   d. proteins and amines
   e. proteins, steroids, and amines

8. Certain hormones can stimulate a _____ .

   a. cellular change
   b. a change in a target organ
   c. a change in a group of organs
   d. a change in the activities of all the cells in the entire body
   e. all of the above

9. The stomach and intestine produce stomach _____ and intestinal _____ , hormones carried by the blood to the _____ glands which then secrete _____ .

   a. enterogastrone, secretin, cholecystokinin, pancreozymin
   b. gastrin, enterogastrone, cholecystokinin, pancreozymin
   c. cholecystokinin, secretin, gastrin, enterogastrone
   d. gastric juice, secretin, gastric, enterogastrone
   e. gastrin, gastrin, gastric, gastric juice

10. Damaged tissues release _____ which dilates the muscles in the walls of the blood vessels, making them more permeable to their contents, enabling more white blood cells and antibodies to move into the damaged area where they can fight infection.

   a. histamine
   b. antihistamine
   c. insulin
   d. glucagon
   e. cortisone

Pancreas, Adrenals and Kidneys, Thyroid

11. The pancreas functions not only in aiding digestion by producing pancreatic digestive enzymes but also contains _____ that produce the hormone _____ which reduces the concentration of _____ in the blood.

   a. islet cells, insulin, glucose
   b. islets of Langerhans, insulin, glucose
   c. adrenal glands, norepinephrine, glycogen
   d. a and b
   e. all of the above

12. An insulin deficiency, known as _____ results in the inability of the liver and muscles to properly control the conversion of glucose to glycogen.

a. diabetes
b. cretinism
c. hypothyroidism
d. hypertensin
e. hyperthyroidism

13. The pancreas releases _____ , which has the opposite effect of _____ , causing the amount of blood glucose to increase.

a. insulin, glycogen
b. insulin, glucagon
c. glycogen, insulin
d. glucagon, insulin
e. none of the above

14. On top of each kidney is an _____ .

a. adrenal gland
b. adrenal cortex
c. adrenal medulla
d. suprarenal gland
e. all of the above

15. Each adrenal gland is composed of _____ and _____ .

a. an outer adrenal cortex, an inner adrenal medulla
b. an outer adrenal medulla, an inner adrenal cortex
c. a thyroid gland, a parathyroid gland
d. an anterior pituitary gland, a posterior pituitary gland
e. none of the above

16. _____ hormones regulate carbohydrate and protein metabolism, salt and water balance, and some sexually related functions as well.

a. cortical
b. pancreatic
c. thyroid
d. gastrointestinal
e. digestive

17. The adrenal medulla secretes _____ .

a. adrenalin
b. epinephrine
c. noradrenaline
d. norepinephrine
e. all of the above

18. The kidneys secrete the protein renin that reacts with a blood protein, forming the hormone _____ that stimulates the contraction of small blood vessels, increasing blood pressure.

    a. hypertensin
    b. angiotonin
    c. erythropoietin
    d. a and b
    e. all of the above

19. An amino acid that has been altered by the addition of four iodine atoms and is the primary hormone produced by the thyroid gland is

    _____ .

    a. thriiodothyronin
    b. throcalcitonin
    c. thyroxin
    d. hypothyroid
    e. hyperthyroid

20. Iodine is necessary for proper thyroid function. With insufficient amounts in the diet the condition _____ may develop, resulting in a decrease in energy.

    a. hyperthyroidism
    b. hypothyroidism
    c. diabetes
    d. iodinemia
    e. parathyroidism

Parathyroids, Thymus, Pituitary, Hypothalamus

21. Parathormone is the hormone produced by the parathyroids, which regulates the _____ balance between the blood and other tissues.

    a. potassium-phosphate
    b. calcium-sodium
    c. potassium-sodium
    d. sodium-phosphate
    e. calcium-phosphate

22. Hyperparathyroidism leads to _____ .

    a. a deficiency of calcium
    b. nervous twitches
    c. spasms
    d. convulsions
    e. the demineralization of bone tissue

23. Hypoparathyroidism is or results in ———————— .

    a. a deficiency of calcium
    b. nervous twitches
    c. spasms
    d. convulsions
    e. all of the above

24. The thymus is composed of ———————— .

    a. tightly packed lymphocytes
    b. lymphocytes held in place by fibrous tissue
    c. T cells
    d. all of the above
    e. none of the above

25. The hormone that stimulates female mammary glands to produce milk is
———————— .

    a. adrenocorticotrophic hormone
    b. luteinizing hormone
    c. progesterone
    d. prolactin
    e. follicle-stimulating hormone

26. The corpus luteum in the ovary secretes ———————— .

    a. adrenocorticotrophic hormone
    b. progesterone
    c. antidiuretic hormone
    d. oxytocin
    e. vasopressin

27. The following hormone stimulates the kidneys to absorb more water,
returning it to the bloodstream, thereby decreasing the urine volume:

    a. antidiuretic hormone
    b. progesterone
    c. oxytocin
    d. luteinizing hormone
    e. thyrotrophic hormone

28. Contraction of the uterus is stimulated by ———————— .

    a. antidiuretic hormone
    b. vasopressin
    c. oxytocin
    d. melanocyte-stimulating hormone
    e. growth hormone

29. The hypothalamus receives nervous impulses from the body's sense organs and then responds by secreting hormone-like substances, known as _____ , into the blood.

a. anticoagulants
b. blood thinners
c. ammonium ions
d. releasing factors
e. all of the above

Ovaries and Testes

30. The menstrual cycle and ovarian cycle are controlled by follicle-stimulating-hormone releasing factor and luteinizing-hormone releasing factor, both of which originate in the _____ .

a. anterior pituitary
b. posterior pituitary
c. infundibulum
d. hypophysis
e. hypothalamus

31. Along with estrogens, _____ prepares the endometrium (the internal layer of the uterus) for implantation of an embryo, and it helps prepare the breasts for milk secretion.

a. progesterone
b. follicle-stimulating-hormone releasing factors
c. luteinizing-hormone releasing factor
d. vasopressin
e. oxytocin

32. _____ are involved in the development and maintenance of female reproductive structures such as the uterus and breasts, as well as being involved in fat distribution and deposition, voice pitch, broadening of the pelvis, and hair growth.

a. estrogens
b. follicle-stimulating-hormone releasing factors
c. luteinizing-hormone releasing factors
d. thyrotrophic hormones
e. all of the above

33. _____ stimulates the seminiferous tubules to begin spermato-genesis.

a. follicle-stimulating hormone
b. testosterone
c. relaxin
d. estrogen
e. luteinizing hormone

34. Just prior to birth, _____ stimulates the descent of the testes into the scrotum.

    a. follicle-stimulating hormone
    b. testosterone
    c. relaxin
    d. estrogen
    e. luteinizing hormone

ANSWERS

1. e	10. a	19. c	28. c
2. c	11. d	20. b	29. d
3. a	12. a	21. e	30. e
4. a	13. d	22. e	31. a
5. d	14. e	23. e	32. a
6. e	15. a	24. d	33. a
7. e	16. a	25. d	34. b
8. e	17. e	26. b	
9. e	18. d	27. a	

CHAPTER 9

The Brain and the Nervous System

THE BRAIN

Evolutionarily, the vast majority of organisms above the level of the primitive invertebrate phyla have evolved sense organs in the anterior portion (front end) of their bodies that respond to the high concentration of incoming information there. The most anterior ganglion, or enlarged, organized, integrative mass of nervous tissue, is called the **brain.** It is the brain that is responsible for processing most of the incoming information.

In many species of invertebrates, the brain is not much larger than the other ganglia located along the rest of the longitudinal nerve cords (see Figure 9.1). The brain of an invertebrate usually has considerably less dominance over the rest of the nervous system, and therefore the body, than is true for a vertebrate brain in the same size category. The brain in most lower vertebrates is not capable of significantly more complex tasks than most invertebrate brains. But the early vertebrate brains reflect evolutionary trends that led to many of the brain developments that have helped distinguish the vertebrates from other groups of organisms.

In the higher invertebrates as well as in the vertebrates, the brain functions in coordination with, or in place of, the many localized, segmented ganglia that are usually little more than a stimulus-and-response apparatus. This large accumulation of nervous tissue receives and transmits sufficiently large amounts of data to give it considerable control over the rest of the organism. The brain also makes it possible for many of these organisms to learn.

MEMBRANES COVERING THE BRAIN

The brain is protected by a system of membranes, the **meninges.** Outside the meninges is the hard, bony covering protecting the head, called the **skull.** The **cranium** is the part of the skull covering the brain. One of the three layers of meninges, the **dura mater,** lies just under the skull; it is tough and fibrous. Just under the dura mater is the meninge that resembles a cobweb, the **arach-**

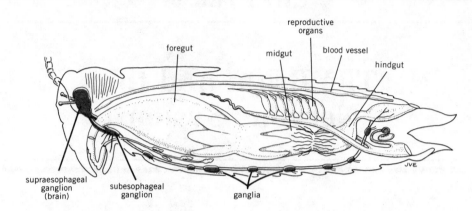

Figure 9.1 *Grasshopper's nervous system (brain and ganglia are shaded).*

noid. And just over the brain is the **pia mater,** which is tightly molded around it. Between the pia mater and the arachnoid is **cerebrospinal fluid**, which acts as a protective cushion, protecting the brain from mechanical injury. Cerebrospinal fluid also fills the central canal that penetrates the entire length of the spinal cord and extends all the way into the brain, where it forms a series of cerebrospinal fluid-filled compartments, known as **ventricles.**

THE VERTEBRATE BRAIN AND ITS EVOLUTION

Even the most primitive vertebrate brain has three principal divisions; each is also found in the advanced vertebrate brain. The most anterior of these three divisions is called the **forebrain.** It is composed of the **olfactory lobes, cerebrum, thalamus, hypothalamus,** and **pituitary.** The **midbrain** connects the forebrain with the hindbrain. It is composed largely of the **optic lobes.** The **hindbrain** consists of the **cerebellum** and the **medulla oblongata.** These parts of the brain are described below, and they are illustrated both on page 140 (Figure 8.3) and on page 154 (Figure 9.2).

Forebrain

The forebrain includes two olfactory lobes (or bulbs) that are always associated with the sense of smell. The posterior part of the forebrain consists of three other main structures. The thalamus is the major sensory integrative area in the forebrain of lower vertebrates. In higher vertebrates, it also integrates some of this information, but these functions have largely, through evolutionary history, become relegated to the cerebrum. The cerebrum, a major part of the forebrain, is the main center for controlling sensory and motor responses, as well as memory, speech, and most factors associated with intelligence. Underneath the cerebrum, near the thalamus, is the hypothalamus, which controls visceral functions such as blood pressure, body temperature, hostility, hunger, pain, pleasure, reproductive behavior, thirst, and water balance.

Midbrain

During evolutionary history, in several of the more advanced lineages, the forebrain increased in relative size and importance. Accordingly, the midbrain decreased in relative size and importance (see Figure 9.2). The most important parts of the midbrain are the specialized areas known as the **optic lobes.** These are the visual centers connected to the eyes by the **optic nerves**.

Hindbrain

The anterior portion of the hindbrain became enlarged and specialized as the cerebellum, which controls balance, equilibrium, and muscular coordination. The ventral portion of the hindbrain, the medulla oblongata, became increasingly specialized as the center of control for such visceral functions as heartbeat and breathing. This is the part of the brain that connects the nerve tracts from the spinal cord to the rest of the brain.

For the most part, **gray matter** consists of cell bodies and synapses. In fish, gray matter is primarily involved in relaying information from the olfactory lobes to the brain. The synapses connecting these neurons act as little more than relays, moving about the neuronal impulses without much integration. Amphibians have more gray matter, indicating that their cerebrum functions less as a simple conduit and more as an integrater of incoming information.

Concomitant with this expansion of internal gray matter, the gray matter moved from the inside part of the brain to the surface, where it is called the **cerebral cortex.** In amphibians and many reptiles, this surface layer of the brain is involved in smell. In some cases, its role expanded to the point where it may have become involved in the control of emotions. Certain advanced reptiles developed an additional component to the cortex, the **neocortex,** which expanded in primitive mammals into a covering over most of the forebrain. The neocortex of more advanced mammals increased in size and became folded, or convoluted, which increased its surface area. As this occurred, the ancestral (olfactory) part of the cortex was relegated to a more internal position in the forebrain.

Both birds and mammals evolved from reptiles, but birds evolved from reptiles without a primitive neocortex. The result has been that modern birds do not have the large, convoluted cerebral cortex found in higher mammals. Since it was another part of the bird cerebrum that grew in size, it is thought that this distinctly different origin of most of the modern cortex of birds and higher mammals explains the differences in their behavior. Birds rely more on innate responses. To a greater extent, mammals appear to develop much of their behavior depending on their experiences. However, when dealing with behavior, there is considerable overlap between and among different groups of organisms.

Originally the midbrain was important as a coordinating center. Later the thalamus area of the forebrain took over much of this function. Eventually the neocortex of higher mammals preempted much of the midbrain and thalamus control, relegating the midbrain to little more than a link between the forebrain

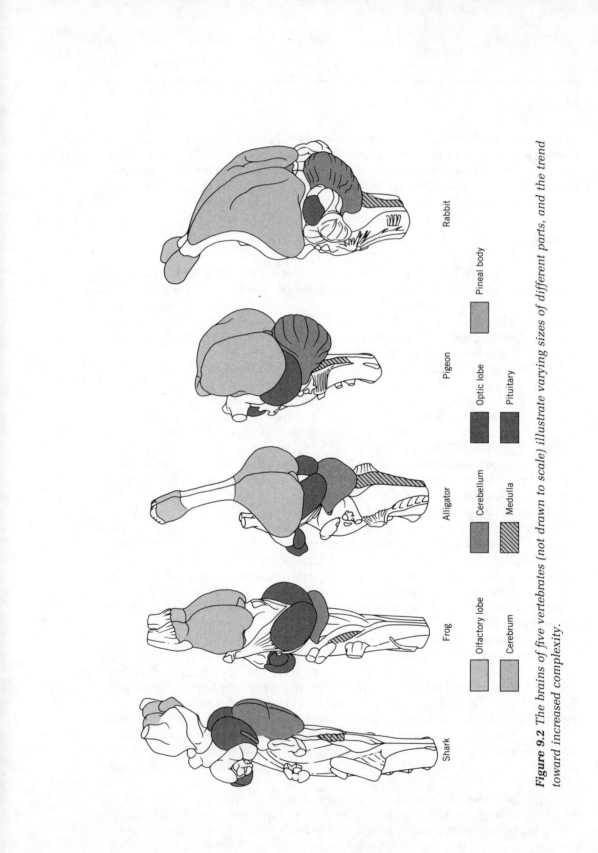

Figure 9.2 The brains of five vertebrates (not drawn to scale) illustrate varying sizes of different parts, and the trend toward increased complexity.

Shark Frog Alligator Pigeon Rabbit

Olfactory lobe Cerebellum Optic lobe Pineal body

Cerebrum Medulla Pituitary

and the hindbrain. The midbrain of modern advanced vertebrates does have some control over several reflexes and minor eye functions, and there is a degree of interaction with emotional responses.

Even though researchers have made great progress in understanding which parts of the brain are involved with specific functions, we still have very little understanding of how the brain accomplishes each of these functions. For instance, although we might know which region of the brain contains certain memories, we cannot yet say with any degree of certainty how memories are made, nor can we say what they are made of or, for that matter, how they are retrieved and integrated.

THE NERVOUS SYSTEM

Neurons are among the most excitable cells in the body. As a group, they respond to a wide range of electrical, chemical, thermal, or mechanical stimuli, transmitting messages to one another, to muscles, and to endocrine organs (hormone-secreting glands). Together, all the neurons and their supporting cells (**glial cells**) compose the **central nervous system**.

Neurons do not exist in sponges (phylum Porifera) or more primitive organisms. The first neurons appear among the coelenterates (phylum Cnidaria), which include jellyfish, hydra, and anemones. Of all living organisms, coelenterates have the simplest nervous arrangement, with only two types of nerve cells: **receptor-conductor cells** (those that respond to the stimuli and pass it on) and **effector cells** (those that contract when the stimulus reaches them). They have none of the alternative types that allow for the increased flexibility of response typical of higher organisms.

As the neurons in a nervous system increase in number, so does the complexity of behavioral responses an animal can have. Since one neuron communicates with nearby neurons, which in turn communicate with other neurons, the total number of possible neuron connections increases exponentially as the total number of neurons increases.

A roundworm (phylum Nematoda) is an organism that moves very little. It has only about 160 neurons. The leech (phylum Annelida), slightly more mobile, has about 13,000 nerve cells. An octopus (phylum Cephalopoda), which has considerable control over its movements and behavior, has over 1 billion neurons. And humans (phylum Chordata) have more than 10 billion neurons.

EVOLUTION OF THE NERVOUS SYSTEM

Nerve Net

Although many groups of lower invertebrates lack a nervous system, cnidarians do possess a simple nervous system termed a **nerve net**. A connected network of neurons without any apparent central control, the nerve net depends on the strength of the original stimulus to transmit a generalized reaction throughout the other neurons. The result may be contraction of other cells that later relax, leading to varying responses that enable simple organisms to maintain rather complex life histories.

Directed Movement

The **radially symmetric** cnidarians were successful, but radial symmetry turned out to be an evolutionary dead end. The flatworms (phylum Platyhelminthes), however, represent a significant advance in neural organization. Flatworms have a distinct top and bottom, front and back, head and tail. Rather than being sessile (attached to the substrate) or drifting about at random, flatworms control the direction of their movement; they have **directed movement.**

Among those species groups with directed movement, natural selection has favored the clustering of neurons in the anterior region, where the incoming information may be processed before being passed on to other neurons. Such neuronal clusters are known as **ganglia** (singular: ganglion).

The evolutionary trend toward the construction of animals with a body axis and directed movement (and away from the basically spherical) led to the neuronal development of the anterior region. This directed and lateral arrangement is termed **bilateral symmetry.** The anterior region of such organisms is the head. The large ganglion in the head that maintains considerable control over much of the entire nervous system, and therefore over much of the body, is often referred to as the brain. Such organisms are said to have a centralized nervous system.

NEURONS

Neurons are the only cells that transmit signals or nervous impulses; glial cells appear to provide nutrition to the neurons. Neurons are usually only a few micrometers in diameter, and most are quite small, though some extend from the spinal cord to the fingertips. Depending on the type of neuron, the long part is usually called the **axon** and the thicker part is the **cell body**, which contains the cell's nucleus. **Dendrites** are the short, branching projections extending from the cell body. They conduct nervous impulses toward the cell body. Axons usually conduct impulses away from the cell body, although they can also carry impulses toward it. In this case, however, there is no effect on an effector organ or cell. Bundles of neurons are called **nerves.**

Cnidarian nerve fibers are more primitive than the generalized neurons described above; they are not differentiated into dendrites and axons. The impulses are conducted in either direction, moving at random throughout the nerve net.

Neuronal responses include the ability to add up many incoming signals and integrate the information. Together, this is accomplished with three basic types of neurons. **Sensory neurons** carry information about environmental change to such integration centers as the brain or spinal cord. **Interneurons** (or **association neurons**) are the major components of the integration centers. They relay messages from one neuron to another. Most neurons in complex animals fall into this category. **Motor neurons** carry the impulses away from the integration centers to muscles or glands (see Figure 9.3).

Associated with the neurons are different types of glial cells. Together, all the glial cells are known as the **neuroglia,** which account for at least half of the nervous system's volume. Some axons are encircled by one type of glial cell, the **Schwann cell,** which provides nutrition to the neuron (see Figure 9.4). Some Schwann cells' plasma membranes envelop certain axons and nerves. Such membranes are called **myelin sheathing**. This resembles fatty insulation and seems to be related to increasing the rate at which the nervous impulses are conducted along the axons. The **nodes of Ranvier** occur at intervals along such axons. These are constricted junctions where one Schwann cell ends and the next begins.

In the brain and spinal cord, much of the nervous tissue consisting of myelinated axons is called **white matter** because the myelinated axons are whitish in appearance. Much of the brain's nervous tissue, lacking a fatty sheath surrounding the axons, is called **gray matter.**

NERVOUS IMPULSE

Both the endocrine system (hormone-secreting glands and their products) and the nervous system control many of the body's activities by regulating and integrating much of what an organism does throughout its life. One of the major functional differences between the nervous system and the endocrine system is the speed with which the nervous system reacts. A nervous impulse can travel through an entire organism in a fraction of a second, while hormones (which move through the blood) elicit a slower response. The speed at which a nervous impulse can travel through myelinated nerves is about 200 km/sec. Nervous impulses travelling through nonmyelinated nerves travel about half as fast (100 km/sec).

Changes in the physical or chemical environment (i.e., due to motion, sound, light, heat, or chemicals) can be converted into nervous impulses. The environmental change is known as the **stimulus,** and the neuronal response is the **neural impulse.** When part of the nervous system receives a neural impulse, it may respond by sending another impulse to the appropriate effectors. Many effectors are muscles, which respond by contracting. However, there are many other types of effectors, such as photoreceptor cells or glandular cells. In addition, the nervous impulse may reach another neuron, which triggers it to the next neuron, and so on, although such an impulse may eventually dissipate to the point that it can no longer elicit a response from an effector.

A neural impulse is triggered by a change in the neuron's electrical charge, which is the result of rapid movement of certain ions. Like most other living cells, neurons have an asymmetric distribution of ions across their plasma

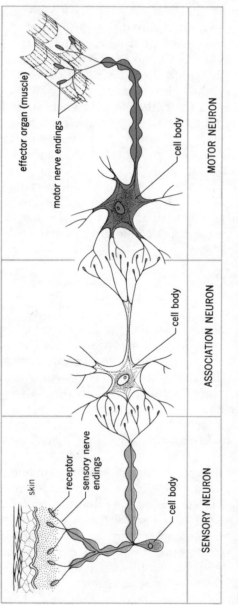

Figure 9.3 *Sensory, association, and motor neurons and their relationships.*

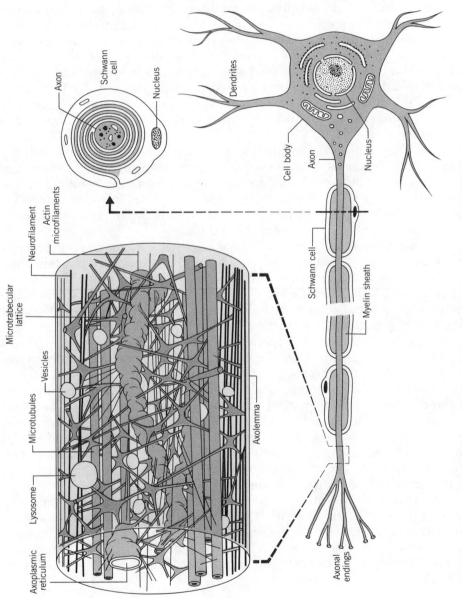

Figure 9.4 A nerve cell and associated Schwann cells with microstructural details.

Schwann
cell

Axon

Nucleus

Dendrites

Cell body

Axon

Nucleus

Schwann cell

Myelin sheath

Axonal
endings

Neurofilament

Actin
microfilaments

Microtrabecular
lattice

Vesicles

Microtubules

Lysosome

Axoplasmic
reticulum

Axolemma

membranes. The interior of a resting neuron (one that is not transmitting an impulse) contains more negatively charged ions than the outside of the cell, where there are more positively charged ions. This uneven distribution of electrical energy, which might be described as an electrical potential difference across the membrane, is usually referred to as **membrane potential,** and is critical to the neuron's ability to transmit an impulse along its entire length.

A resting neuron has a net negative charge inside the axon, primarily from negative chloride ions (Cl^-), and a higher concentration of positively charged sodium ions (Na^+) outside the neuron. Inside the neuron, along with the negative chloride ions, are positively charged potassium ions (K^+). Upon a neuronal stimulus of sufficient strength, a response, or a **nervous impulse** is initiated. The intensity of stimulus required to activate this kind of response is called **threshold**.

At the site where the neuron is stimulated, the membrane initially becomes more permeable to sodium ions, which then rush across the membrane to the inside of the cell, momentarily producing a slightly more positive charge inside the cell relative to the outside. The membrane also becomes a little more positively charged relative to the outside of the cell. This change in membrane potential is called **depolarization.**

After the neuron becomes more permeable to sodium ions, it becomes more permeable to potassium ions. So in the fraction of a second following the influx of sodium ions, potassium ions rush out of the cell. This exit of positively charged ions restores the charge inside the cell as well as that of the membrane to its initial negative charge. Once depolarization has been initiated at one end of a neuron, it passes down the entire length of the neuron. This sequence of events is known as a **nervous impulse.**

This wave of depolarization, also called an **action potential,** rapidly passes down the entire length of the axon. This is why an action potential is termed "all or nothing." Then, immediately following the inrush of sodium ions and the outflow of potassium ions, the sodium ions are pumped back out of the nerve cell and the potassium ions are pumped back in (against their concentration gradient, so energy is required to fuel the active transport). The pump that restores the original ionic balance is the **sodium-potassium pump**. It does this after a very brief **refractory period,** when the neuron can't conduct a neural impulse (see Figures 9.5 and 9.6).

Upon reaching the end of the neuron, which is called an axon ending or axon terminal, the action potential stimulates the release of chemicals known as **neurotransmitters,** which travel across the **synapse** (a short gap) to the next neuron (see Figure 9.7). These chemicals trigger a new depolarization, and then another action potential travels along the next neuron too (see Figure 9.8).

SYNAPSE AND NEUROTRANSMITTERS

The term synapse is used to describe both the junction between a neuron and another neuron and that between a neuron and the cell it acts upon.

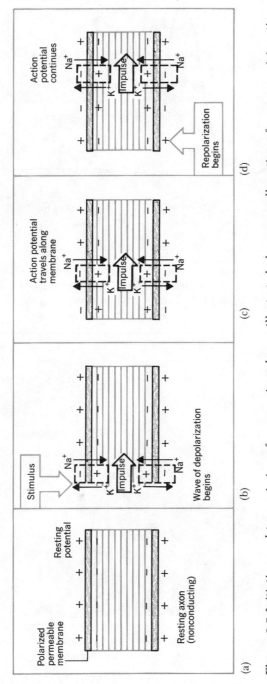

Figure 9.5 *Initiation and transmission of a nerve impulse as illustrated along a small section of a neuron: (a) resting nerve fiber; (b) impulse begins with depolarization of cell membrane as sodium ions (Na^+) move into the cell and potassium ions (K^+) move out; (c) the wave of depolarization moves along the nerve fibers; (d) after the impulse passes, the membrane repolarizes by pumping out the sodium ions.*

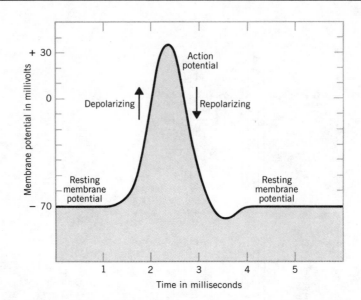

Figure 9.6 *This graph illustrates the change in a nerve fiber's membrane potential when measured at one location along the neuron during a nerve impulse. The movement of sodium ions into the cell depolarizes it, and the movement of potassium ions out of the cell repolarizes it.*

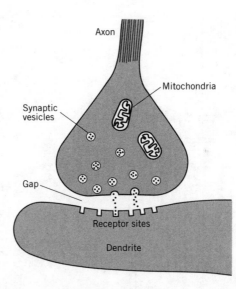

Figure 9.7 *A synapse, the junction between two neurons. Neurotransmitters are released from synaptic vesicles and picked up by receptor sites.*

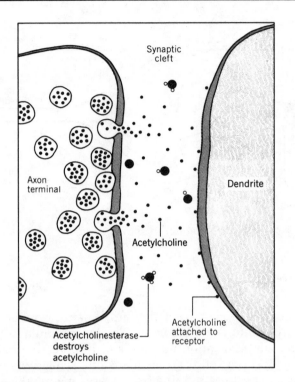

Figure 9.8 *After the neurotransmitter molecules attach to the receptor sites, they are deactivated by specific enzymes.*

The gaps between the cells at the synapse are about 200 Å wide. There are synapses between the ending of an axon and a dendrite, between an axon and a cell body, and sometimes between two axons. Most neurons synapse (*synapse* can also be used as a verb) with a number of neurons, although many also synapse with other cells such as muscles and glands. The axons produce neurotransmitters, which cross the synapse and are picked up at receptor sites on the dendrites of an adjoining neuron.

The difference between inhibition or stimulation depends on the amount and type of neurotransmitter as well as on the type of receptor site. Over 10 different neurotransmitters have been identified to date; these include acetylcholine, dopamine, glutamate, gamma-aminobutyric acid, histamine, noradrenaline, and serotonin. Each affects the response of a neuron. Normally, neurotransmitters are rapidly broken down enzymatically following release. But tranquilizers, caffeine, nerve gas, many insecticides, and curare (the chemical used in poison arrows in South America) can interfere with neurotransmitters—by stimulating or retarding neurotransmitter production, by binding to a receptor site, or by affecting the enzymes that normally destroy the neurotransmitters.

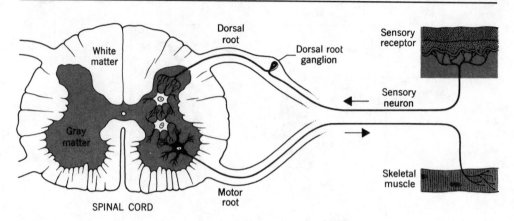

Figure 9.9 *Cross section of the spinal cord showing major features and the route of a reflex arc.*

REFLEX ARCS

A **reflex arc** is usually based on a small group of neurons where the entire neural impulse totally circumvents the brain. The most simple reflex arc contains a sensory neuron, an association neuron, and a motor neuron that synapses with an effector cell. Each reflex arc contains one sensory neuron, sometimes extremely long, such as those running from a large mammal's foot to its spinal cord. The cell bodies of these sensory neurons are always located in ganglia (**dorsal-root ganglia**) just outside the spinal cord (see Figure 9.9). The axons enter the spinal cord dorsally, where they connect (synapse) with several association neurons in the gray matter of the spinal cord. Then they synapse with a motor neuron (also in the gray matter of the spinal cord) and exit the spinal cord ventrally. Figure 9.9 uses a cross section of a spinal cord to illustrate the path of a reflex arc.

Some of the association neurons directly synapse with motor neurons that run back to where the sensation originated. The advantage of a reflex arc is speed. The signal is sent straight back, causing an immediate response, such as a knee-jerk reaction, an eye blink, or the formation of a tear drop. Reflex arcs are not the result of any conscious control. However, it is possible for some association neurons to synapse with other association neurons that pass up the spinal cord to the brain, where the information is relayed to other centers. There it may then be possible to consciously inhibit part of the reflex, or add to it.

Still other reflexes may be far more complex than the simple knee-jerk reaction. Two such examples are breathing and the control of one's heart beat. Though largely involuntary, it is possible to exert a considerable amount of conscious control over these behaviors.

All the nerves connecting with the spinal cord contain sensory and motor neurons and are called mixed nerves. Not all the nerves intercept with the spinal cord. Humans have 12 pairs of nerves that directly connect with the

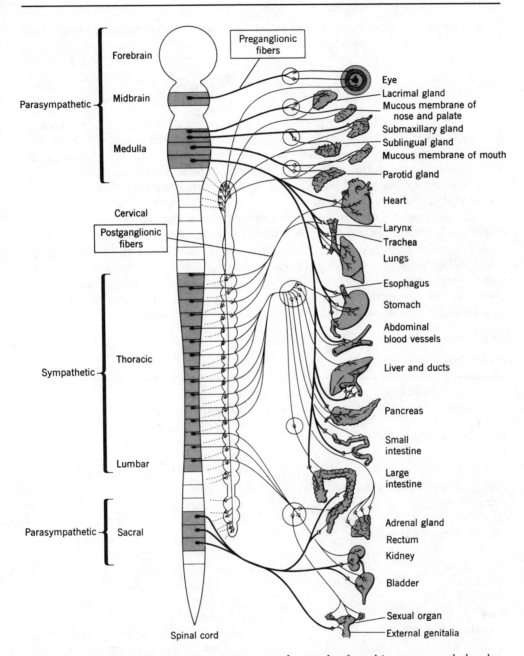

Figure 9.10 *Autonomic nerves innervate smooth muscles found in organs and glands. Most organs receive innervation from sympathetic and parasympathetic portions of the nervous system.*

brain. These are known as the **cranial nerves,** some of which contain just sensory or just motor neurons, and others are mixed.

ORGANIZATION OF THE NERVOUS SYSTEM

The brain and spinal cord compose the central nervous system. The **somatic nervous system** conducts nervous impulses that have already been processed, away from the central nervous system to the skeletal muscle tissue. The somatic nervous system is under voluntary control. All the parts of the nervous system, excluding the brain and spinal cord, are collectively known as the **peripheral nervous system.**

The **autonomic nervous system** consists of the nerves that carry nervous impulses from the central nervous system to the heart (cardiac muscles), to the muscles in the digestive system (smooth muscles), and to the glands (see Figure 9.10). All of these muscles and glands contract and function involuntarily. The autonomic nervous system is subdivided into two parts, the **sympathetic** and **parasympathetic systems**. These function in opposition to one another; the first inhibits organs, while the latter usually excites organs.

Key Terms

action potential	medulla oblongata
arachnoid	membrane potential
association neuron	meninges
autonomic nervous system	midbrain
axon	motor neuron
bilateral symmetry	myelin sheathing
brain	neocortex
cell body	nerve
central nervous system	nerve net
cerebellum	nervous impulse
cerebral cortex	neural impulse
cerebrospinal fluid	neuron
cerebrum	neurotransmitter
cranial nerves	nodes of Ranvier
dendrites	olfactory lobes
depolarization	optic lobes
directed movement	optic nerves
dorsal-root ganglia	parasympathetic nervous system
dura mater	peripheral nervous system
effector cells	pia mater
forebrain	pituitary
ganglia	radial symmetry
glial cells	receptor
gray matter	receptor-conductor cells
hindbrain	reflex arc
hypothalamus	refractory period

Schwann cell
sensory neuron
skull
sodium-potassium pump
somatic nervous system
stimulus

sympathetic nervous system
synapse
thalamus
threshold
ventricles
white matter

Chapter 9 Self-Test

QUESTIONS TO THINK ABOUT

1. What protective layers envelope the vertebrate brain?
2. How are vertebrate brains similar in basic construction?
3. What do the terms forebrain, midbrain, and hindbrain refer to?
4. What is the difference between gray and white matter?
5. What are three basic types of neurons?
6. Define myelin sheathing and explain its function.
7. What is the all-or-nothing principle?
8. Shortly after being released, what normally happens to neurotransmitters?
9. How does a nervous impulse pass down a neuron?

MULTIPLE-CHOICE QUESTIONS

The Brain

1. The brain is protected by a membranous system known as the _____
   _____ .

   a. ventricles
   b. olfactory sheaths
   c. oblongata
   d. meninges
   d. medullas

2. The _____ is the part of the skull that covers only the brain.

   a. dura mater
   b. arachnoid
   c. pia mater
   d. cerebrospinal fluid
   e. cranium

3. The spinal cord has a central canal that extends into the brain, becoming a series of hollow compartments called _____ , which are filled with cerebrospinal fluid.

    a. ventricles
    b. dura mater
    c. pia mater
    d. meninges
    e. thalamus

4. The first of the meninges that lies just under the skull, and is tough and fibrous, is the _____ .

    a. dura mater
    b. pia mater
    c. ventricle
    d. thalamus
    e. gray matter

5. Between the pia mater and the arachnoid is the _____ that bathes the entire region, providing a cushion to protect the brain from mechanical injury.

    a. albumin
    b. ovalbumin
    c. lymph
    d. cerebrospinal fluid
    e. protoplasm

6. The part of the brain that consists of the olfactory bulbs, cerebrum, thalamus, hypothalamus, and pituitary is the _____ .

    a. forebrain
    b. midbrain
    c. hindbrain
    d. medulla oblongata
    e. cerebellum

7. The cerebellum and the medulla oblongata are part of the _____ .

    a. forebrain
    b. midbrain
    c. hindbrain
    d. thalamus
    e. hypothalamus

8. The _____ is the major sensory integrative area in the forebrain of lower vertebrates.

    a. olfactory bulbs
    b. medulla oblongata
    c. cerebellum
    d. thalamus
    e. hypothalamus

9. The _____ controls visceral functions such as blood pressure, body temperature, hostility, hunger, pain, pleasure, reproductive behavior, thirst, and water balance.

  a. thalamus
  b. hypothalamus
  c. medulla oblongata
  d. olfactory bulbs
  e. cerebrum

10. The _____ controls balance, equilibrium, and muscular coordination.

  a. thalamus
  b. hypothalamus
  c. gray matter
  d. olfactory bulbs
  e. cerebellum

11. The part of the brain that consists of cell bodies and synapses is known as the _____ .

  a. optic nerves
  b. optic lobes
  c. olfactory bulbs
  d. white matter
  e. gray matter

The Nervous System

12. Together, all the neurons and their supporting cells comprise the _____ .

  a. glial cells
  b. myelin sheathing
  c. nerve net
  d. nervous system
  e. central nervous system

13. Among members of the phylum Cnidaria, the connected network of neurons that creates the nervous system lacking central control is known as a _____ .

  a. central nervous system
  b. ganglion
  c. brain
  d. axon
  e. nerve net

14. Flatworms have a top and bottom, a front and back, a head and tail; this type of body construction is known as _____ .

    a. radial symmetry
    b. lateral symmetry
    c. bilateral symmetry
    d. trilateral symmetry
    e. asymmetry

15. All groups of animals above the evolutionary level of sponges have a _____ .

    a. radially symmetrical arrangement
    b. bilaterally symmetrical arrangement
    c. nerve net
    d. tentacle
    e. nervous system

16. Receptor cells _____ stimuli.

    a. receive
    b. conduct
    c. speed up
    d. slow down
    e. circulate

17. Conductor cells are specialized for _____ stimuli.

    a. conducting
    b. stopping
    c. slowing down
    d. speeding up
    e. associating

18. Effector cells are usually _____ or _____ .

    a. muscle, bone
    b. bone, gland
    c. gland, muscle
    d. none of the above
    e. all of the above

19. As the entire nervous system became more complex in terms of its increased flexibility of response, there was a trend toward _____ .

    a. specialization
    b. cephalization
    c. minimization
    d. maximization
    e. a and b

20. The part of a neuron that contains the cell's nucleus is the _____ _____ .

    a. axon
    b. dendrite
    c. cell body
    d. glial cell
    e. myelin sheath

21. _____ are usually short, branching projections extending from the cell body.

    a. perikaryons
    b. dendrites
    c. cell bodies
    d. Schwann cells
    e. axons

22. In terms of total volume, all the glial cells account for about _____ _____ of the nervous system.

    a. 10%
    b. 25%
    c. 50%
    d. 75%
    e. 100%

23. All of the _____ are known as the neuroglia.

    a. glial cells
    b. neurons
    c. axons
    d. cell bodies
    e. dendrites

24. Some of the axons are encircled by _____ that provide nutrition and perhaps other forms of support.

    a. Schwann cells
    b. cell bodies
    c. perikaryons
    d. nerves
    e. nodes of Ranvier

25. The _____ resemble(s) a coiled, fatty insulation around certain axons.

    a. myelin sheath
    b. nodes of Ranvier
    c. ganglia
    d. sensory neurons
    e. association neurons

26. Myelinated axons are _____ in appearance.

    a. greenish
    b. reddish
    c. bluish
    d. grayish
    e. whitish

27. The junction between two or more neurons is called a _____ .

    a. neural impulse
    b. neurotransmitter
    c. synapse
    d. dorsal-root ganglion
    e. nervous junction

28. The movement of the electrical impulse across the synapse requires specific chemicals known as _____ .

    a. neural impulses
    b. neurotransmitters
    c. synapse jumpers
    d. dorsal-root ganglia
    e. nervous junction chemicals

29. Bundles of individual axons are called _____ .

    a. synapses
    b. sensory neurons
    c. nerves
    d. gray matter
    e. association neurons

30. Each reflex arc contains one sensory neuron that has its cell body located just outside the spinal cord in the _____ .

    a. motor neuron
    b. association neuron
    c. spinal cord
    d. dorsal-root ganglion
    e. gray matter

31. All the parts of the nervous system, excluding the brain and spinal cord are collectively known as the _____ .

    a. afferent system
    b. efferent system
    c. peripheral nervous system
    d. somatic nervous system
    e. autonomic nervous system

32. The _____ consists of nerves that carry nervous impulses from the central nervous system to the smooth muscles, the heart muscle, and to the glands.

    a. afferent system
    b. efferent system
    c. peripheral nervous system
    d. somatic nervous system
    e. autonomic nervous system

33. The parasympathetic system usually _____ an organ.

    a. inhibits
    b. excites
    c. carries lymph to
    d. carries lymph from
    e. a and b

34. The sympathetic system usually _____ the particular organ.

    a. inhibits
    b. excites
    c. carries lymph to
    d. carries lymph from
    e. a and b

ANSWERS

1. d	10. e	19. e	28. b
2. e	11. e	20. c	29. c
3. a	12. d	21. b	30. d
4. a	13. e	22. c	31. c
5. d	14. c	23. a	32. e
6. a	15. e	24. a	33. a
7. c	16. a	25. a	34. b
8. d	17. a	26. e	
9. b	18. c	27. c	

CHAPTER 10

Bones and Muscles

BONES

Muscles and **bones** work together. The bones make up the **skeletal system**, which provides structural support, sites for muscle attachment, and organ protection. **Osseous tissue**, or bone, as it is more often called, consists of cells and **collagen fibers** interspersed in a matrix of intercellular material containing **calcium phosphate** and **calcium carbonate**, which are responsible for hardness. Together, these substances account for two-thirds of the weight of bones, while the collagen fibers, which reinforce the tissue, account for the other third.

In addition to bone, another important connective tissue in most skeletal systems is **cartilage**, which, unlike bone, is both firm and flexible. Bone is usually considerably harder and more brittle. Most sharks and rays have skeletal systems composed of all cartilage and no bone. Some other "primitive" groups of fish have less bone than cartilage in their skeletal systems. In most other vertebrates, however, cartilage is located only where firmness and flexibility are needed, such as in joints, nose, ears, larynx, and trachea. During the development of the skeletal system of these vertebrates, the embryos begin with cartilaginous skeletons. Gradually most of the cartilage is replaced by true bone.

Depending on the construction of the particular bony tissue, it can range in consistency from being completely spongy to being very compact. The spongy bone contains many spaces filled with **marrow** (which is either composed of fat or involved in the production of blood cells. In the case of relatively lighter animals such as birds, the spaces may be filled with air sacs. Compact bony tissue is thicker and usually involved in support. Such bones can resist considerable weight and stress.

Compact bones are penetrated by blood vessels and nerves through small narrow openings, some of which are known as **Haversian canals**, whose microscopic structure is identified by the characteristic concentric rings of bony tissue surrounding them. These rings are composed of cells that were involved in producing the bony tissue. Spongy bone doesn't contain Haversian systems, nor does cartilage. Materials are exchanged through the blood vessels and bone cells that penetrate the Haversian canals. This is the only way for materials to move to and from the cells living throughout bony tissue.

The main bones in the human body, from head to toe, are as follows. The fused bones creating the **cranium** compose the **skull**; the lower **teeth** are located in the **mandible**, or jaw. The collar bone is the **clavicle**. The "wings" in the upper back are called **scapulas**. The bone connecting all the **ribs** in the middle of the chest is the **sternum**. The ribs are connected in the back to the **vertebral column** (backbone), which is composed of **vertebrae**. The vertebrae in the neck are called **cervical vertebrae; thoracic vertebrae** articulate with the ribs; **lumbar vertebrae** descend from the thoracic vertebrae to the pelvis; and together, the fused bones in the **pelvis** compose the **sacrum**. The tail is composed of **caudal vertebrae**. In humans, the "tail" is called the **coccyx** (see Figure 10.1).

The bone in the upper arm is the **humerus**, and the two bones in each lower arm are the **radius** and **ulna**. The wrist bones are called **carpals**. At the base of the fingers, located within the part of the hand known as the palm, are the **metacarpals**, and the smaller bones extending out to the fingertips are the **phalanges**.

The largest bones in the body are the **femurs**, the thigh bones that connect the upper leg with the pelvis. The distal end of each femur is attached to the lower leg. The upper and lower legs meet at the **knee** covering which is the kneecap, or **patella**. Each lower leg has two long bones, the **tibia** and **fibula**. The little bones in the ankle area are the **tarsals**; then come the **metatarsals**. The rest of the bones extending to the toe's tips are, like the fingers, called **phalanges** (see Figure 10.2). Note similarities between the human and pigeon skeleton, Figure 10.3.

Some bones are held together with fused, immovable joints, such as the **sutures** located between several skull bones. Other joints are movable and are held together with **ligaments**, the flexible tissues that connect bones and cartilage. Similarly, **tendons** are flexible tissues connecting muscles to bones. The end of a muscle attached to the bone nearest to the axis of the body (proximal) is known as the **origin**. The muscle end attached to the farther bone (distal), such as in the hand or foot, is known as the **insertion**. In Figure 10.4, for example, the biceps originates on both the humerus and the scapula, and it inserts on the radius.

The movements of different parts of the body all depend on muscular contraction, on the location of the origins and insertions, and on the type of joint involved. Muscles usually work antagonistically. That is, when one group of muscles contracts, it will pull part of the body one way. Alternately, when the antagonistic group of muscles contracts, it will pull the same body part in the other direction. The biceps and triceps in Figure 10.4, for example, form an antagonistic muscle pair.

MUSCLES

Three types of muscles have been recognized in vertebrates. These are **smooth**, or **visceral muscle; skeletal muscle;** and **cardiac**, or **heart muscle**. Smooth muscles line internal organs such as the intestines and bladder. They also line

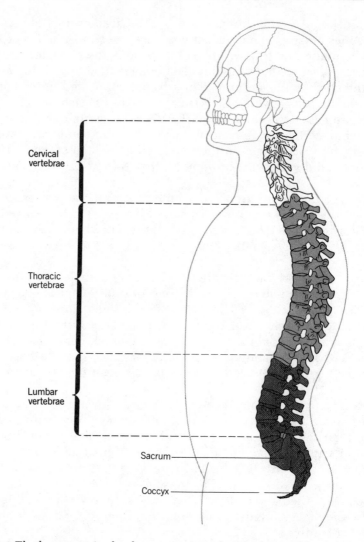

Cervical
vertebrae

Thoracic
vertebrae

Lumbar
vertebrae

Sacrum

Coccyx

Figure 10.1 *The human spinal column consists of 33 bones. Within the spinal column is the spinal cord, branching from which are 33 pairs of spinal nerves that emerge from between the bones.*

the walls of the arteries and veins, and many ducts and tubes found throughout the body. On a cellular level, smooth muscles differ from the other two types in that each cell (or muscle fiber) contains one nucleus and is long, thin, and pointed at both ends. For the most part, the smooth muscles contract involuntarily (they are innervated by the autonomic nervous system). Together, these muscle fibers form thin broad sheets.

Responding to conscious control, the skeletal muscles are commonly called the voluntary muscles (they are innervated by the somatic nervous system). Skeletal muscles move the arms and legs, back, face, jaw, and eyes, as well as many other parts of the body. Skeletal muscle cells (fibers) are **coeno-**

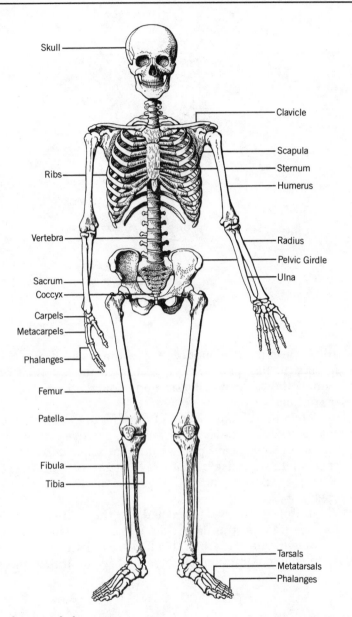

Figure 10.2 The human skeleton.

cytic (contain many nuclei) and, when viewed microscopically, are crossed by many thin dark lines, which is why they are called **striated muscles**. Together, many skeletal muscle fibers form **bundles**, which are wrapped in connective tissue to form muscles.

Like skeletal muscle cells, cardiac muscle fibers have striations. They are also multinucleate, but they are innervated by the autonomic nervous system. The heart has a **pacemaker**, which spontaneously begins each heartbeat. This specialized area located in the wall of the right atrium (see Figure 12.4, page

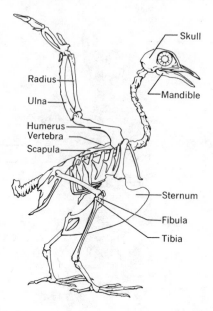

Figure 10.3 The pigeon skeleton.

207) is called the **sinoatrial node.** Once the heart begins to contract, the impulse spreads to the node lying near the atrium between both ventricles, which is called the **atrioventricular node**; once stimulated, it initiates the ventricular contraction.

Unlike vertebrate muscle, all insect muscles are striated, even those lining the internal organs. Many other invertebrates, however, have both smooth and striated muscle, and some have only smooth muscle. One of the key differences between each of these muscle types is that striated muscle contracts very rapidly but, unlike smooth muscle, cannot be held in the contracted position for very long.

Smooth muscle cells are connected to (thus innervated by) two nerve fibers. When one fiber is stimulated, the muscle cell contracts, and, when the other is stimulated, the muscle cell relaxes. Sometimes, as is the case with involuntary muscle contraction, such as that in the throat when swallowing,

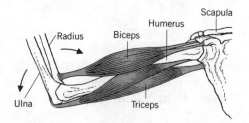

Figure 10.4 With balance sensors located in muscles and tendons, the antagonistic action of muscle pairs controls the tensing and releasing of different muscles. Together, this antagonistic action enables an organism to control its body movements.

the contractions constituting the wave of **peristalsis** may occur without direct nervous stimulation.

MUSCLE CONTRACTION

Like nerve cells, muscle cells contract either entirely or not at all. When a muscle receives a nervous stimulus, the actual response, or how strongly the muscle contracts and how much work it can do when contracting, depends on the number of muscle cells stimulated. That, in turn, depends on the strength of the initial stimulus. Not all of the individual muscle fibers are alike; some respond to stronger stimuli than others. By increasing the stimulus, more muscle fibers contract until all contract, and that is the **maximal stimulus**. After reaching a maximal stimulus, any increase in stimulus will not elicit a stronger muscular contraction.

When a muscle is stimulated, a certain base level of electricity is necessary to produce a simple twitch and, after this contraction, it takes a brief interval for the muscle to contract and then relax before it can contract again. The time taken to contract is the **contraction period**, and the time taken to relax before the muscle can contract again is the **relaxation period**. The amount of time from when the initial stimulus is administered until the contraction begins is called the **latent period**. Together, the latent, contraction, and relaxation periods constitute a single simple **muscle twitch**.

If a muscle is not allowed to relax completely before being stimulated again, the next contraction stimulated by the same electrical input elicits a stronger response. If one continually stimulates the muscle, eliciting stronger and stronger contractions until the maximum contraction is reached, the period of increased contractions is called **summation**, and the leveling off to one sustained contraction is called **tetanus**. Afterward the muscle **fatigues**. Figure 10.5 shows a **kymograph**, which records the intensity of muscle contractions over time.

The energy required for muscle contraction is fueled by adenosine triphosphate (ATP), which is stored in the muscles until needed. ATP is a triple-phosphorylated organic compound that functions as "energy currency" in most organisms. The oxygen found in the muscles is stored in **myoglobin**, a compound quite similar to hemoglobin. The harder muscles have to work, the more oxygen they consume. Without enough available oxygen, working muscles continue to contract, deriving energy through a different biochemical pathway. This alternative, known as **anaerobic respiration**, causes a **lactic acid** build-up, via **fermentation**, which can be poisonous. Anaerobic respiration causes what is termed an **oxygen debt**, which means that after strenuous muscular activity, one breathes very deeply to acquire the needed oxygen to convert this potentially dangerous lactic acid to **glycogen**, which is a useful carbohydrate. (The concepts mentioned in this paragraph are explained in Chapter 5).

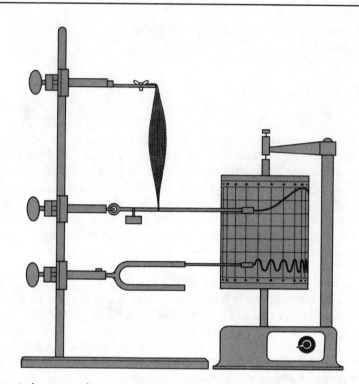

Figure 10.5 A kymograph is used in laboratory experiments to study muscle contractions. A pen, which is attached to a system of levers, records muscle contractions. The muscle can be stimulated with known amounts of electricity for known durations. Each muscle contraction is then recorded on a revolving drum. The muscle can be treated with chemical stimulants or depressants, and the muscle's reaction recorded. A timing device or a tuning fork creates a series of values for comparison and calibration.

SLIDING-FILAMENT THEORY

The contractile elements that make up most of the muscles' bulk consist of two proteins, **actin** and **myosin**. Alone, neither protein will contract, but together, they form an **actomyosin complex** that, in the presence of ATP, will contract. An individual muscle fiber is composed of many long thin **myofibrils**, each of which looks like a long ribbon with alternating light and dark bands. The wide light bands are the **I-bands**; they are composed of actin. In the middle of each I-band is a dark line called the **Z-line**. The broad dark bands, the **A-bands**, are composed of myosin. Each has a lighter **H-zone** through the middle. The unit from one Z-line to the next, along a single myofibril, is called a **sarcomere** (see Figure 10.6).

When the muscle contracts, the actin and myosin slide together with the light and dark areas overlapping. When the muscles relax, the proteins slide apart again (see Figure 10.7).

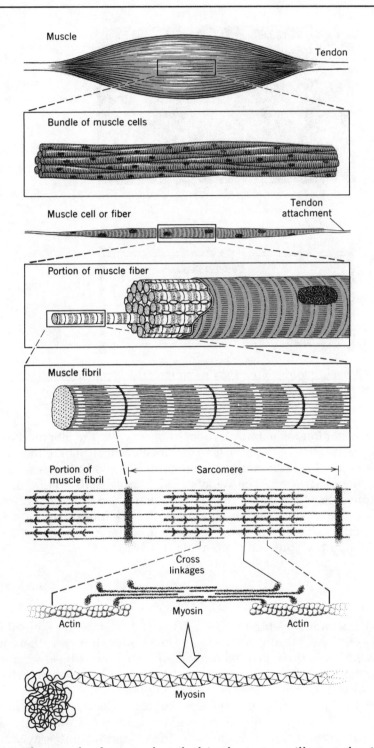

Muscle

Tendon

Bundle of muscle cells

Muscle cell or fiber

Tendon attachment

Portion of muscle fiber

Muscle fibril

Portion of muscle fibril

Sarcomere

Cross linkages

Actin

Myosin

Actin

Myosin

Figure 10.6 *The muscle elements described in the text are illustrated, with this breakdown of a muscle's organization.*

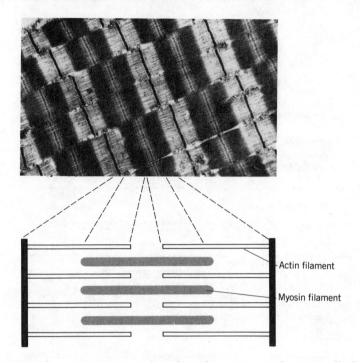

Figure 10.7 *Muscle contraction results when the protein polymers called actin and myosin slide together. Actin filaments are thinner than myosin filaments.*

The contraction of a muscle fiber depends on the depolarization of a polarized, resting cell. A nerve carries an electrical signal that stimulates a transmitter at the neuromuscular junction. This stimulates a momentary reduction of polarization. The depolarized muscle cell admits an inflow of **calcium ions** (Ca^{++}). The repolarization phase depends on the outflow of **potassium ions** (K^+) and, at the same time, the calcium pump moves calcium ions back out of the cell. This wave of depolarization spreads across the nerve cell, stimulating the contractile elements to slide together.

EXOSKELETON

So far only the **endoskeleton** has been discussed. That is the skeletal structure of animals with cartilage or bones inside the body. However, there are many organisms, far more than the total number of vertebrates (those organisms with an internal bony and/or cartilaginous skeleton), that have tough, hard external skeletons, or **exoskeletons**; some are even jointed.

Arthropods (including insects) have an exoskeleton, with the hard covering outside the body and all muscles and organs located internally. The hard outer covering is noncellular. Secreted by the epidermis (outer layer of skin), it prevents excessive water loss, acts as armor protecting the creature, and provides sites for muscle attachment that can withstand the pressure and weight of muscle contraction. These organisms have movable joints that bend when their antagonistic muscles are contracted.

Key Terms

A-bands
actin
actomyosin complex
anaerobic respiration
atrioventricular node
bones
bundles
calcium carbonate
calcium ions
calcium phosphate
cardiac muscle
carpals
cartilage
caudal vertebrae
cervical vertebrae
clavicle
coccyx
coenocytic
collagen fibers
contraction period
cranium
endoskeleton
exoskeleton
fatigue
femurs
fermentation
fibula
glycogen
Haversian canals
heart muscle
humerus
H-zone
I-bands
insertion
knee
kymograph
lactic acid
latent period
ligaments
lumbar vertebrae
mandible
marrow
maximal stimulus

metacarpals
metatarsals
muscle fiber
muscle fibril (myofibril)
muscle twitch
myofibril
myoglobin
myosin
origin
osseous tissue
oxygen debt
pacemaker
patella
peristalsis
phalanges
potassium ion
radius
relaxation period
ribs
sacrum
sarcomere
scapula
sinoatrial node
skeletal muscle
skeletal system
skull
smooth muscle
sternum
striated muscle
summation
sutures
tarsals
teeth
tendon
tetanus
thoracic vertebrae
tibia
ulna
vertebrae
vertebral column
visceral muscle
Z-line

Chapter 10 Self-Test

QUESTIONS TO THINK ABOUT

1. What substances are found in skeletal systems? Explain their function.
2. How do muscles work antagonistically?
3. The vertebral column is made of which types of vertebrae? Where are they located?
4. What are the three types of muscle found among vertebrates? Give an example of each.
5. What is the function of the pacemaker?
6. Describe muscle contraction and relaxation.
7. Explain the fundamentals of the sliding-filament theory.

MULTIPLE-CHOICE QUESTIONS

Skeletal and Muscular Systems

1. Which of the following helps protect organs, provides sites for muscle attachment, and lends structural support?

   a. skeletal system
   b. tendons
   c. ligaments
   d. heart muscle
   e. bone marrow

2. In addition to bone, another kind of connective tissue comprising many skeletal systems, which is firm, though not as hard and brittle as bone, is _____ .

   a. calcium phosphate
   b. calcium carbonate
   c. cartilage
   d. marrow
   e. Haversian canals

3. Blood vessels and nerves penetrate compact bones through small narrow openings, some of which are known as _____ .

   a. ligaments
   b. tendons
   c. Haversian canals
   d. tarsals
   e. phalanges

4. Some bones are held together with fused joints, which are immovable, such as the _____ located between several skull bones.

   a. tendons
   b. ligaments
   c. sutures
   d. cartilage
   e. ribs

5. Some individual muscle fibers respond to stronger stimuli than others, so the stimulus that makes all the muscle fibers contract is known as the _____ .

   a. peristalsis
   b. summation
   c. contraction period
   d. maximal stimulus
   e. initial stimulus

6. Sometimes, as is the case with involuntary muscle contraction, such as that in the throat when swallowing, the contractions comprising the wave of _____ may occur without direct nervous stimulation.

   a. sinoatrial contraction
   b. maximal stimulus
   c. peristalsis
   d. latent contraction
   e. contraction period

7. The time a muscle takes to relax before the muscle can contract again is the _____ .

   a. latent period
   b. contraction period
   c. fermentation period
   d. lactic acid period
   e. relaxation period

8. The time a muscle takes to contract is the _____ .

   a. contraction period
   b. latent period
   c. fermentation period
   d. lactic acid period
   e. relaxation period

9. The instrument used to record the intensity of muscle contractions over time is a(n) _____ .

   a. electrocardiogram
   b. electroencephalogram
   c. kymograph
   d. X-ray machine
   e. ultrasound machine

10. Oxygen found in muscles is stored in ——————— , a compound quite similar to hemoglobin.

    a. plasma
    b. white blood cells
    c. kymoglobin
    d. myoglobin
    e. cartilage

11. Without available oxygen, working muscles continue to contract, deriving energy through a biochemical pathway known as ——————— .

    a. anaerobic respiration
    b. aerobic respiration
    c. photosynthesis
    d. summation
    e. relaxation

12. Anaerobic respiration, via fermentation, causes ——————— .

    a. summation
    b. latency
    c. kymography
    d. lactic acid build-up
    e. tetanus

13. After strenuous muscular activity, one breathes very deeply to acquire the needed oxygen to convert the potentially dangerous lactic acid to

    ——————— .

    a. actin
    b. myosin
    c. actomyosin
    d. myofibrils
    e. glycogen

14. The contractile elements that make up most of the muscles' bulk consist of two proteins that alone will not contract but together form a(n) ——————— that in the presence of ATP will contract.

    a. actin
    b. myosin
    c. actomyosin complex
    d. myoglobin
    e. hemoglobin

15. An individual muscle fiber is composed of many long ——————— .

    a. muscles
    b. tendons
    c. ligaments
    d. bones
    e. myofibrils

16. The unit from one Z-line to the next, along a single myofibril, is called a(n) ——————— .

    a. sarcomere
    b. A-band
    c. H-zone
    d. I-band
    e. actin

17. A depolarized muscle cell admits ——————— .

    a. calcium ions
    b. potassium ions
    c. magnesium ions
    d. sodium ions
    e. chloride ions

18. A tough or hard external skeleton is called a(n) ——————— .

    a. endoskeleton
    b. exoskeleton
    c. arthropod
    d. chitin
    e. cartilaginous skeleton

ANSWERS

1. a	6. c	11. a	16. a
2. c	7. e	12. d	17. a
3. c	8. a	13. e	18. b
4. c	9. c	14. c	
5. d	10. d	15. e	

CHAPTER 11

Internal Transport: Plants and Invertebrates

Chemical reactions of all living cells occur in an aqueous environment, which explains why the materials within metabolically active cells are suspended or dissolved in water. In addition to the aqueous internal environment the cells of many plants and animals are bathed in a nutritious extracellular environment, eliminating the need for specialized systems to move substances to and from these cells. The requirements of larger organisms, however, may differ considerably. Most large plants and animals have an internal transportation system that suits their particular life history. In general, the larger or more mobile the animal, the more complex and faster moving is the internal transport capability.

Both plants and animals have tubular transport systems for distribution of materials within their bodies. Most plants have two major pathways for internal transport: the **phloem**, which carries carbohydrates, and the **xylem**, which carries water and ions. Many animals have a system of arteries, veins, and lymphatic vessels, which carry blood and lymph, the protein-containing fluid that escapes from the blood capillaries. These transport systems are described in more detail below, after a discussion of some organisms that do without such special transport systems.

ORGANISMS WITHOUT INTERNAL TRANSPORT SYSTEMS

The internal contents of many small, relatively simple organisms can be moved around without the complex of internal structures such as tubes, vessels, and special mechanisms observed in most higher organisms. Single-celled organisms such as bacteria and protists rely on **diffusion** as one of their major transport systems. It may or may not be fast, but, with such small distances involved, the random distribution resulting from diffusion of most microscopic substances throughout the internal space available seems to be an important, if not a primary, internal transport mechanism. Even in multicellular organisms, diffusion plays an important role at the cellular level.

While diffusion accounts for much of the movement of fluids and solutes within cells, **intercellular diffusion** accounts for movement between cells.

This important mechanism helps move material from one cell to another in multicellular organisms. In plants, intercellular diffusion may be facilitated by **plasmodesmata**, the strands of protoplasm that penetrate cell walls and connect the cytoplasm of adjacent cells (see Chapter 2).

Often, diffusion is a very slow process, one that is supplemented by other mechanisms. One process, known as **cytoplasmic streaming**, has been observed in many cells; the cytoplasm flows along what appears to be a definite route throughout the cell, moving substances many times faster than would otherwise be possible. **Food vacuoles** often move throughout a cell, distributing digested material to different parts of the cytoplasm.

Some plants, such as the bryophytes—which include the liverworts, hornworts, and mosses—lack vascular tissues, the efficient long-distance internal transport systems that are responsible for moving fluids throughout the bodies of higher plants. The absence of such tissues has probably limited the size attained by most of these plants for two reasons. First, when the vascular tissue xylem is present, it can also function as a major supportive tissue in vascular plants. Second, without an efficient internal transport system, the plants are unable to move materials rapidly around great distances, thus restricting the overall size.

There are other internal transport mechanisms possessed by multicellular organisms. These represent major alternatives to those found among the larger, more dominant plant and animal groups. Jellyfish, hydra, and planaria lack true circulatory systems, but they have other mechanisms that free them from relying completely on diffusion and intracellular transport. For instance, the **gastrovascular cavity** of the hydra penetrates each tentacle, enabling food particles to be absorbed by each tentacle cell. Planaria have branched gastrovascular cavities extending throughout their body; this system moves food particles to all their cells without the aid of a circulatory system.

TRANSPORT IN PLANTS

The term vascular, when referring to specific tissues, has to do with those tissues concerned with tubular internal transport, such as the xylem and phloem in plants. The evolution of these tissues has enabled plants to develop greater heights, more specialized parts, and more highly integrated functions. It is widely believed that the successful exploitation of terrestrial environments by plants followed the evolution of these complex internal transport systems.

The vascular tissue is continuous throughout the plant, extending through the roots, stems, twigs, and leaves, as well as other parts of the organism. Here, for the sake of simplicity, the major components of this system will be considered individually.

STEMS

The outermost tissue layer of the stem of **herbaceous** plants, those plants without woody parts, is called the **epidermis**. Inside this is the **cortex**, which sometimes is divided into two layers: the one just under the epidermis, called the **collenchyma**, and the innermost, the **parenchyma**. Just internal to the cortex is the vascular tissue.

Vascular tissue in some plants is arranged in bundles. In others it occurs as a continuous layer, like a cylinder, lying inside the cortex with the **pith** filling the innermost part of the stem. The pith is a storage area. Those plants with vascular tissue arranged in bundles tend to have no clear distinction between the cortex and the pith. Rather, the tissue throughout the interior part of the stem appears to be quite homogenous.

In other types of plants, the phloem lies outside the xylem, with a layer of tissue in between the two that is primarily involved with producing new cells through mitotic cell division. This middle layer is the **lateral meristematic tissue**, or **cambium**. After the initial growth phase of many plants, the cambium of herbaceous species ceases to produce more phloem and xylem cells. However, in other species, the cambium reactivates at the beginning of the next growing season and continues to produce more phloem cells on the outside, and xylem cells on the inside. In some plants (e.g., trees), the old phloem can be seen flaking off. The **bark** includes all the tissues outside the **vascular cambium**. On the inside, the old xylem continually forms additional rings of tissue that gradually add to the stem's or trunk's diameter. These are the growth rings, which appear in cross section when a tree is sawed down. As years go by, more xylem gets laid down, more phloem continually flakes off, and the stem increases in thickness; eventually the bulk of the stem of the older plant is made primarily of xylem. This tissue is commonly called **wood**. Some plants also have a **cork cambium** located outside the **cortex**, which divides mitotically to create the most external tissue, known as **cork** (see Figure 11.1).

CONDUCTING TISSUES IN PLANTS

Phloem

As stated earlier, phloem carries carbohydrates, the products of photosynthesis, generally in the form of sucrose, from the leaves to the nonphotosynthetic parts of the plant. The conducting cells of the phloem are the **sieve elements**, which are joined together in vertical columns, creating the **sieve tubes**, each of which is joined at the end, at the **sieve plates**. Where the sieve elements meet, the sieve plates are perforated, facilitating the movement of intracellular contents from cell to cell.

Next to the sieve elements are companion cells that seem to be involved in maintaining the sieve elements. The companion cells are living when they mature, though they lose many of their organelles, including the nucleus, which disintegrates. For the sieve elements to remain functional, they have to retain an intact cell membrane, as well as their watery cytoplasm. The absence of most organelles in the phloem cells may promote the mass movement of water and solutes through the sieve tubes.

Most of the movement of the contents in the sieve tubes occurs from the regions where plant food is produced, to the sites of food storage or utiliza-

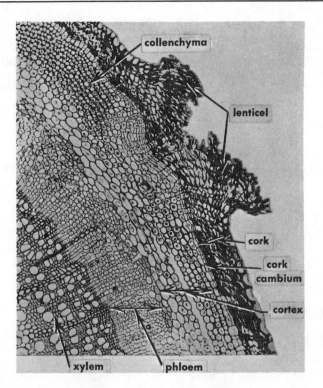

Figure 11.1 *Going from the inside to the outside of the stem, the tissues shown in this elderberry stem cross section are xylem, phloem, cortex, cork cambium, and cork. The vascular cambium consists of the first two layers of thin-walled cells on the phloem side of the xylem.*

tion, where the solutes and water move from the sieve tubes into the adjacent cells and into the xylem. In addition to carbohydrates, organic nitrogen compounds move through the phloem, being conducted not only downward, but, depending on where the nutrients are needed, laterally and upward as well.

Xylem

Xylem is the plant tissue that carries water and ions from the roots to other parts of the plant. It is laid down along the inside of the cambium (lateral meristematic tissue). Unlike phloem, the xylem's conducting cells, the **tracheids** and **vessel elements**, die at maturity and become empty shells consisting of primary and secondary cell walls without a cell membrane, cytoplasm, nucleus, or any organelles. (The additional types of xylem cells are rarely mentioned in most introductory texts.)

Tracheids are long, narrow cells tapering at the ends where they come in contact and overlap with each other. The first tracheids to be laid down by the plant are characterized by **annular** and/or **spiral** structures. In addition to having **meristematic cells** (capable of active cell division and differentiation into specialized tissues) that develop into phloem and xylem cells, the vascular cambium also has meristematic cells that develop into a system of cells oriented at a right angle to the stem's axis. These structures, known as **rays**, carry food from the sieve tubes radially into the cambium and xylem. And sometimes the rays store large quantities of carbohydrates.

ROOTS

Most **roots** are storage organs, holding minerals and carbohydrates for future use, but they also function as a holdfast, stabilizing the plant (see Figure 11.2). And, of course, they have the vascular tissues, xylem and phloem. Much of the absorption of water and minerals takes place through the **root hairs**, which are thin outgrowths of epidermal cells. They lack the waxy cuticle that protects other thin structures, such as leaves. In many species, a symbiotic relationship exists between the plant's root hairs and the filamentous elements of specific fungi, which form the mycelium. The product of such a close association is known as **mycorrhizae**.

Root Pressure

There seem to be several mechanisms that account for the upward movement of fluids and solutes from one cell to the next. **Root pressure** is responsible for some of the movement of water across the root tissues. It appears to be a function, in part, of simple diffusion along a concentration gradient. However, more seems to be involved. Some other pressure besides simple diffusion is present since the water in the xylem is more concentrated than that in the protoplasm of the **endodermal tissue** (the tissue surrounding the vascular cylinder). In other words, if simple diffusion alone were involved, the water probably wouldn't just move into the xylem, but also out of it. In addition, one would expect the downward hydrostatic pressure exerted by the column of water standing in the xylem would be expected to force the water out of the vascular tissue and into the surrounding root tissues. Instead, as it actually happens, water goes in and up the xylem.

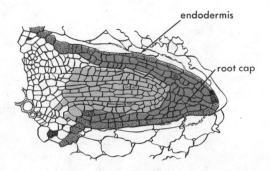

Figure 11.2 *At the start, a young wheat root begins growing by pushing out through the cells of the cortex.*

The force exerted by the roots disappears when the roots are killed or deprived of oxygen. It has been found that the respiratory production of ATP is necessary to provide the energy to create a solute gradient that is at least partially responsible for the movement of water through the root. In addition to the osmotic pressure and the use of semipermeable membranes, facilitated diffusion and active transport may also be involved.

Cohesion Theory

Another hypothesis, the **cohesion theory**, sometimes called the **transpiration theory**, states that water is pulled from above. According to this concept, water is lost through **transpiration**, the escape of water vapor from the aerial parts of the plant, and is replaced by the fluids in adjoining cells, creating an osmotic gradient that draws water from the xylem in the veins of the leaf to the adjacent tissues. The removal of water from the xylem for other uses, such as in photosynthesis, growth, or other metabolic processes, would also contribute to the draw, bringing more water up the xylem. This theory is strengthened by knowledge of the great cohesive forces that exist between individual water molecules.

This system is not, however, run simply by suction, which usually results from the removal of air, but by the removal of water, creating a pull that affects all the water below. Suction alone could not pull so much water to such heights.

The inside of the thin tubes may be made of a material to which water molecules adhere. This, in conjunction with the annular rings and the spiral secondary cell walls, probably accounts for part of the necessary architecture that allows water to rise, in some cases as high as hundreds of feet, defying most projections of what would otherwise be possible. In addition, the internal design may also reduce the water's downward hydrostatic pressure.

Another factor is the considerable weight of the atmosphere resting on the ground. The atmosphere creates enough pressure to account for pushing water up the xylem some 32 feet (almost 10 meters). Together, atmospheric pressure, combined with the effects related to transpiration, the adhesive forces that prevent the water column from breaking under the pressure of gravity, capillary action that draws water up the thin tubes resulting from the adhesion of the water molecules to the inside of the xylem, the cohesion of the water molecules to each other, and the intricate design of the vessel's interior seem to account for a good portion of the upward movement of water observed in the xylem.

MOVEMENT OF INORGANIC SOLUTES

Through the xylem, such **inorganic ions** as calcium, phosphorus, and sulfur are transported up from the roots to the leaves. Although phosphorus ions are capable of moving up through the xylem and down through the phloem, calcium ions are not nearly so mobile in the phloem and cannot be transferred from dying leaves to newer ones. Plants therefore have to obtain a steady supply of new calcium ions from the soil.

Different substances move through the sieve tubes at different speeds; sugar moves through some phloem 40,000 times faster than when diffusing through a liquid. The direction of movement through any given sieve tube reverses periodically, sometimes with the contents in neighboring sieve tubes moving in the opposite direction. Obviously, the transport through living phloem cells is unlike that through the dead xylem cells.

It has been suggested that materials are carried through the phloem cells by cytoplasmic streaming. Materials move from phloem cell to phloem cell through the sieve plates by diffusion; active transport may also be involved. Then they are carried to the other end of the cell by the cytoplasmic streaming, and then diffuse out again. Cytoplasmic streaming involves movement in both directions. This could explain the suggestion that some materials move in one direction, while others move in the opposite direction, through any given cell, though this has yet to be proven. Facilitated diffusion may also be involved.

Another hypothesis involves **pressure flow** of water and solutes through the sieve tubes along a turgor-pressure gradient. Known as the **mass-flow hypothesis**, it states that cells such as those in the leaves contain high concentrations of osmotically active substances such as sugar. So water diffuses into these and adjacent cells, raising the internal pressure, or turgor. This pressure forces substances from one cell to the next, pushing solutes further up the plant. Since the sugars are being used up in the metabolically active areas, such as those which are actively growing, as well as in the cells that are sequestering the sugars in storage organs, osmotic concentrations are lowered in the nearby sieve tubes. Therefore, they lose water and turgor pressure. This results in a mass flow of the contents of the sieve tubes from the regions under high pressure. Such regions include leaves, where sugars are being made, and storage organs such as the roots, from which sugars exit during the spring when they are in transit to where they are needed. The contents then move to areas of low pressure, such as those regions where reserves are being rapidly consumed—where cells are growing or where materials are being put into storage. As presented, this system depends on the massive uptake of water at one end because of high osmotic concentrations, and on the massive loss of water at the other end, because of the low osmotic concentrations that are due to the constant depletion of sugar and other solutes.

Key Terms

annular secondary cell walls	endodermal tissue
bark	epidermis
cambium	food vacuoles
cohesion theory	gastrovascular cavity
collenchyma	herbaceous
cork	inorganic ions
cork cambium	intercellular diffusion
cortex	lateral meristematic tissue
cytoplasmic streaming	mass-flow hypothesis
diffusion	meristematic cells

mycelium
mycorrhizae
parenchyma
phloem
pith
plasmodesmata
pressure flow
rays
root
root hairs
root pressure
sieve elements

sieve plates
sieve tubes
spiral secondary cell walls
tracheids
transpiration
transpiration theory
vascular cambium
vascular rays
vessel elements
wood
xylem

Chapter 11 Self-Test

QUESTIONS TO THINK ABOUT

1. Describe the conducting tissues of plants.
2. What are the theories concerning how materials are conducted throughout plants?
3. What is the function of the roots, and how do they work?

MULTIPLE-CHOICE QUESTIONS

Plant Internal Transport

1. Most plants have the following major pathways for internal transport:
   a. arteries and veins
   b. arteries, veins, and lymphatic vessels
   c. xylem, phloem, arteries, and veins
   d. trachea, dorsal longitudinal vessels, xylem, and phloem
   e. xylem and phloem

2. Many animals have the following major pathways for internal transport:
   a. xylem and phloem
   b. arteries, veins, xylem, and phloem
   c. arteries, xylem, and phloem
   d. arteries, veins, and lymphatic vessels
   e. lymphatic vessels, xylem, and phloem

3. Single-celled organisms such as bacteria and protists rely on _____ _____ as one of their major transport systems.
   a. diffusion
   b. xylem and phloem
   c. arteries and veins
   d. cilia and flagella
   e. all of the above

4. In plants, intercellular diffusion may be facilitated by _____.
   a. food vacuoles
   b. microtubules
   c. cytoplasmic streaming
   d. plasmodesmata
   e. all of the above

5. The following often move throughout a cell, distributing digested material to different parts of the cytoplasm:
   a. endoplasmic reticulum
   b. microtubules
   c. food vacuoles
   d. cytoplasmic streaming
   e. all of the above

6. Some plants, such as the bryophytes, which include the liverworts, hornworts, and mosses, _____.
   a. have vascular tissues
   b. lack vascular tissues
   c. have gastrovascular cavities
   d. have a circulatory system
   e. have xylem and phloem

7. When referring to specific plant tissues, the following term concerns those involved with internal transport:
   a. cortex
   b. collenchyma
   c. parenchyma
   d. epidermis
   e. vascular

8. It is thought that the evolution of vascular tissue enabled plants to develop _____.
   a. greater height
   b. more specialized parts
   c. more highly integrated functions
   d. more successfully on land
   e. all of the above

9. When plants have vascular tissue, it is usually found in the _____.
   a. roots
   b. stems
   c. twigs
   d. leaves
   e. all of the above

10. The outermost tissue layer of an herbaceous plant's stem is called the
    _____ .

    a. epidermis
    b. cortex
    c. collenchyma
    d. parenchyma
    e. pith

11. Directly inside the vascular cambium is the _____ .

    a. cortex
    b. pith
    c. xylem
    d. collenchyma
    e. parenchyma

12. Just outside the cork cambium is the _____ .

    a. cork
    b. collenchyma
    c. parenchyma
    d. vascular tissue
    e. pith

13. Those tissues concerned with tubular internal transport, such as xylem
    and phloem, are called _____ .

    a. epidermis
    b. collenchyma
    c. parenchyma
    d. pith
    e. vascular

14. In between the phloem and the xylem lies the _____ .

    a. lateral meristematic tissue
    b. vascular cambium
    c. apical meristematic tissue
    d. a and b
    e. a, b, and c

15. In some plants, especially trees, the old _____ can be seen
    flaking off, as the old bark.

    a. xylem and phloem
    b. phloem, cortex, and cork
    c. xylem
    d. a and b
    e. all of the above

Conducting Tissues

16. Next to the sieve elements are _____ that seem to be involved in maintaining the sieve elements.

    a. sieve tube
    b. sieve plates
    c. companion cells
    d. all of the above
    e. none of the above

17. _____ carries carbohydrates, usually in the form of sucrose, from the leaves to the nonphotosynthetic parts of the plant.

    a. xylem
    b. phloem
    c. arteries
    d. veins
    e. lymphatics

18. _____ is the plant tissue that carries water and ions from the roots to other parts of the plant.

    a. xylem
    b. phloem
    c. sieve tube
    d. sieve plate
    e. companion cell

19. _____ is the vascular tissue laid down on the inside of the cambium (lateral meristematic tissue).

    a. xylem
    b. phloem
    c. sieve tube
    d. sieve plate
    e. companion cell

20. Unlike the phloem, the xylem's conducting cells, the _____ and _____ , die at maturity and become empty shells consisting of primary and secondary cell walls, without a cell membrane, cytoplasm, nucleus, or any organelles.

    a. sieve tubes and sieve elements
    b. sieve plates and sieve tubes
    c. companion cells and sieve elements
    d. vessel elements and sieve tubes
    e. tracheids and vessels elements

21. _____ are long, narrow cells with tapering ends that overlap where they come in contact with each other. The first to be laid down are characterized by annular and/or spiral secondary cell walls that are involved in moving fluids up through the xylem tissue.

   a. tracheids
   b. vessel elements
   c. companion cells
   d. sieve tubes
   e. sieve elements

22. _____ that grow later in the plant's life have numerous pits which allow water and dissolved substances to move from cell to cell.

   a. tracheids
   b. vessel elements
   c. companion cells
   d. sieve tubes
   e. sieve elements

23. _____ are more specialized than tracheids, from which they probably evolved, and are found primarily in the more advanced, flowering plants.

   a. tracheids
   b. vessel cells
   c. companion cells
   d. sieve tubes
   e. sieve elements

24. In the xylem are clusters of parenchyma cells forming _____ , that instead of running laterally up and down the plant, as do the other internal transport cells, run radially, and facilitate lateral movement of materials.

   a. tracheids
   b. vessels cells
   c. rays
   d. sieve elements
   e. companion cells

25. Much of the absorption of water and minerals takes place through the _____ , which are thin outgrowths of epidermal cells.

   a. rays
   b. tracheids
   c. companion cells
   d. sieve tubes
   e. root hairs

Movement of Fluids Through Vascular Tissues

26. When the roots are killed or are deprived of oxygen, the force that seems to be pushing fluids up, known as _____ , is no longer present.
    a. root pressure
    b. cohesion pressure
    c. transpiration pressure
    d. endodermal pressure
    e. atmospheric pressure

27. The _____ theory states that water is pulled from above.
    a. cohesion
    b. root
    c. transpiration
    d. a and b
    e. a and c

28. According to the cohesion theory, water is lost through _____ , and it is replaced by the fluids in adjoining cells, creating an osmotic gradient that draws water from the xylem in the veins of the leaf to the adjacent tissues.
    a. cohesion
    b. transpiration
    c. insulation
    d. circulation
    e. pressurization

29. _____ can only account for pushing water up xylem about 32 feet.
    a. cohesion pressure
    b. transpiration pressure
    c. endodermal pressure
    d. root pressure
    e. atmospheric pressure

ANSWERS

1. e	9. e	17. b	25. e
2. d	10. a	18. a	26. b
3. a	11. a	19. a	27. e
4. d	12. d	20. e	28. b
5. c	13. d	21. a	29. e
6. b	14. d	22. a	
7. e	15. a	23. b	
8. e	16. c	24. c	

CHAPTER 12

Circulatory System

Like most plants and invertebrates, vertebrates also have a system that delivers nutrients, minerals, and dissolved gases to and from each of their cells, and then carries off the metabolites such as carbon dioxide and nitrogenous wastes. Methods of internal exchange used by plants and some invertebrates were discussed in the preceding chapter. This chapter is primarily concerned with the circulatory systems of insects and vertebrates.

Higher animals usually have a **closed circulatory system** through which their blood circulates, meaning it is repeatedly cycled throughout the body. The blood is the medium that flows through the tubes (arteries, veins, and capillaries), carrying specialized cells, proteins, nutrients, dissolved gases, and metabolites throughout the body. The force maintaining the flow is usually provided by the pumping action of a hollow muscular organ; sometimes there is more than one of these organs, known as **hearts**, that receive blood in one end and pump it out the other.

The one-way pumping action is often accompanied by a series of one-way **valves** that help regulate the movement of blood through a designated route. An **open circulatory system** is characterized by large open sinuses through which the blood flows. Two major groups of animals with open circulatory systems are the mollusks (snails, clams, squid, and octopuses) and arthropods (spiders, ticks, mites, centipedes, millipedes, insects, crabs, and lobsters). In addition to a large open sinus, open circulatory systems also have hearts and vessels.

Animals as varied as annelids (which include earthworms) and vertebrates (fish, amphibians, reptiles, birds, and mammals) have vessels, without an open sinus; such circulatory systems are termed closed. Before discussing closed circulatory systems, open circulatory systems will be described, using insects as the major example.

INSECTS: OPEN CIRCULATORY SYSTEMS

By definition, virtually any physiological system common to the insects is evolutionarily successful since insects account for over 80 percent of all living species. Since insects are in many respects the most important group of invertebrates, and since their circulatory system is so different from ours, it is important to evaluate what makes their system so effective.

Although one of the main functions of the closed circulatory system is to deliver oxygen to all the tissues, that is not always the primary role of the open circulatory system. Unlike other organisms, insects have a **tracheal system**, a network of hollow tubes that delivers air through pores in the body wall, or exoskeleton, to the cells throughout the insect's interior. For this reason, the insect's circulatory system does not deal with oxygen exchange (see Figure 12.1).

Without lungs, and with tracheae, the insect's circulatory system has evolved from the ancestral system, which was probably closed, to something new that utilizes an open internal body cavity, the **haemocoel**.

INSECT BLOOD

Insect blood, **haemolymph**, is composed of water, plasma, minerals, nutrients, waste products, and blood cells, or **haemocytes**. Since oxygen is brought to the tissues through small pores in the exoskeleton, it is clear that the insect's circulatory system differs considerably from that of animals with lungs. Because these systems are so different, it is necessary to explain how the insect's circulatory system works, and why its design differs so much from the vertebrate's closed circulatory system.

The blood enters a **dorsal longitudinal vessel**, which is sometimes called the heart, through openings called the **ostia**, when the vessel is relaxed. When the heart contracts, the blood exits through the anterior opening into the head region. Beyond the dorsal longitudinal vessel, insects have no arteries, veins, or capillaries. The blood simply circulates throughout the haemocoel (see Figure 12.2).

The blood, composed of haemocytes, is bathed in the plasma that carries the nutritive materials secreted from the digestive system to tissues where they are metabolized. Haemocytes appear to be involved in the formation of connective tissue and are also concerned with the activation of certain glands. They have also been shown to assist in coagulation and epidermal regeneration.

Metabolic wastes and dead cells have to go somewhere. Since they cannot be passed out through the tracheae, they are passed into the haemocoel, where some are consumed by haemocytes through **phagocytosis**. The larger particles, such as metazoan, or multicellular, parasites, cannot be consumed by the haemocytes, so they are **encapsulated**, meaning enclosed in a sheath or capsule, and may be stored or eventually passed out through the alimentary canal.

Those waste products that cannot be consumed, recycled, or encapsulated and stored somewhere enter the alimentary canal through the **Malpighian tubules**. These are blind sacs that absorb water, salt, and nitrogenous wastes at one end, where they are bathed in blood. At the other end of the sacs

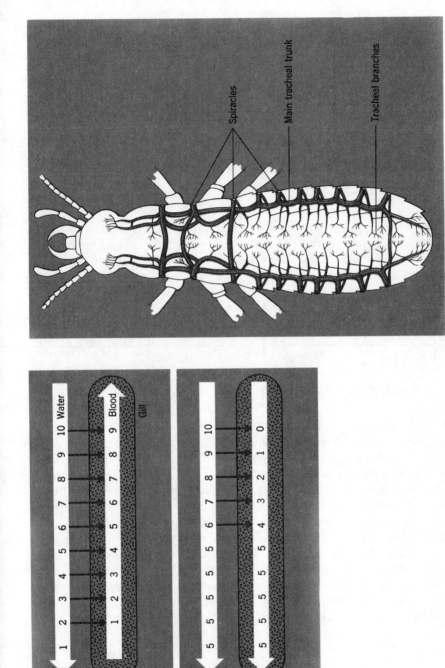

Figure 12.1 On the left is the counter current exchange mechanism as it works in fish gills. When water and oxygen flow into the gills in opposite directions (top) oxygen will move into the blood. The exchange would be much less efficient if both were to flow in the same direction (bottom). On the right is the insect respiratory system. Insects don't breathe through their mouths; air enters the spiracles and moves to the tissues through the tracheae.

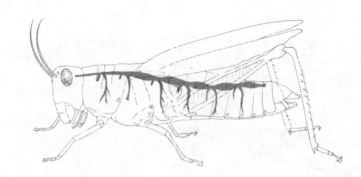

Figure 12.2 *Open circulatory system of a grasshopper illustrating blood flow from the dorsal longitudinal vessel.*

the contents are passed into the hindgut, from which they continue into the rectum. Here water is resorbed and the drier waste, or fecal matter, is excreted.

The blood of many arthropods (spiders, ticks, mites, centipedes, millipedes, and insects) and mollusks (snails, clams, squid, and octopuses) contains an oxygen-carrying pigment called **hemocyanin**, which is like the hemoglobin found in vertebrate blood. Unlike hemoglobin, hemocyanin never occurs in cells but is dissolved in the plasma of those animals that have it. Another difference between these two similar oxygen-carrying substances is that hemocyanin contains copper while hemoglobin contains iron.

CIRCULATION IN VERTEBRATES

Insect hearts and other invertebrate hearts are simpler versions of the more complex hearts that evolved in the vertebrates. In vertebrates, the blood is enclosed within a system composed of a heart and vessels. The heart pumps the blood into large **arteries**, which branch into smaller arteries (**arterioles**), which branch into the smallest blood vessels, the **capillaries** (see Figure 12.3). It is through the thin walls of the capillaries that nutrients, gases, hormones, waste products, and other molecules are exchanged between the blood and the **interstitial fluids** that surround and bathe the body's cells. After passing through the capillaries, the blood passes into small **veins**, or **venules**, which merge into larger veins. Eventually, these return all the blood to the heart. Therefore, this system, which is known as the **cardiovascular system**, circulates blood throughout the body, bringing it to and from the capillaries, where the blood is able to accomplish its primary functions.

In addition to the similarities, there are differences that exist among the major groups of vertebrates (fish, amphibians, reptiles, birds, and mammals). From group to group, there is increasing separation between the left and right sides of the heart. This separation decreases the amount of mixing between the oxygen-rich, or **oxygenated**, blood returning from the gills or the lungs, and the oxygen-poor, or **deoxygenated**, blood returning to the heart from the tissues.

Fish hearts have four internal **chambers** arranged linearly, with one-way **valves** between each chamber. The blood comes in one end and is pumped

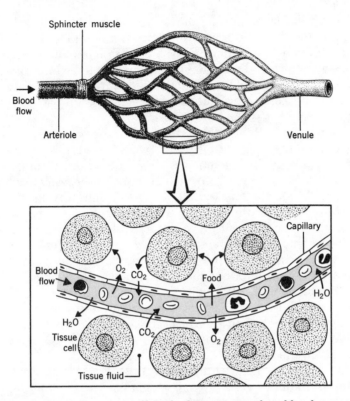

Figure 12.3 *Rate of blood flow to capillary bed (top) is regulated by the constriction or relaxation of the arteriole's sphincter muscle. The exchange of water, nutrients, gases, wastes, and other chemicals occurs between the capillary's blood and the surrounding tissue cells (bottom).*

out the other. There is no mixing of oxygenated and deoxygenated blood in the fish heart because the aerated blood flows straight from the gills to the rest of the body without first returning to the heart (see Figure 12.1 on page 203).

Amphibians have a three-chambered heart, with two atria and one ventricle. The deoxygenated blood flows from the body to the heart, first entering a chamber known as the **right atrium**, which acts as a storage chamber. The thin-walled atrium expands when the blood enters. Then the valves open to allow the blood to flow into the **ventricle**, a thicker chamber whose walls are made of muscle. When the muscle fibers contract, the blood is forced out through the **pulmonary artery** to the **lungs**. From there the oxygenated blood flows back to the heart, entering the storage chamber on the other side, called the **left atrium**. The blood flows back to the muscular ventricle, which then contracts and pumps the oxygenated blood to the rest of the body.

The only problem with this heart design is that there is some mixing of the oxygenated and deoxygenated blood in the ventricle, despite certain compensating devices such as the grooves in a frog's heart that channel blood flow and minimize mixing.

Like the amphibians, reptiles have two atria. But rather than just one ventricle, there are two. These ventricles are not completely divided in most reptiles, although the ventricular division is complete in all bird and mammal hearts. This is why bird and mammal hearts are said to be four-chambered (two atria, two ventricles). It is the four-chambered heart that has enabled only birds and mammals to possess completely separate circulation paths for arterial and venous blood.

The human heart, with a right atrium and right ventricle, as well as a left atrium and left ventricle, essentially has two separate hearts inside one (see Figure 12.4). Both beat simultaneously. Blood circulates through a human in the following way. After flowing through all parts of the body except the lungs, the deoxygenated blood enters the heart via the right atrium. From the right atrium the blood is forced through the **tricuspid valve** into the right ventricle. The right ventricle pumps the deoxygenated blood through the **pulmonary semilunar valve** into the **pulmonary artery**, which divides into two main branches, each entering one of the lungs. In the lungs, the pulmonary arteries branch into arterioles, which connect with the **capillary beds** in the walls of the **alveoli**, the small air sacs throughout the lungs. This is where gas exchange occurs. Upon inhalation, the oxygen passes from the lungs into the thin layer of fluid lining the alveoli, and then it is picked up by the hemoglobin inside the red blood cells. Also at the alveoli, carbon dioxide is released from the blood to the air in the lungs, which is then exhaled.

From the capillaries in the lungs, blood passes into small veins and then into the larger **pulmonary veins**, two of which lead from each lung back to the heart, where they empty into the left atrium. The blood then passes through the **bicuspid valve** (also called the **mitral valve**) and enters the **left ventricle**. The left ventricle contracts and pumps the blood through the **aortic semilunar valve** into the **aorta**, the artery that carries the oxygenated blood out of the heart and to all parts of the body except the lungs (see Figure 12.4).

Oxygenated blood passes through arteries that lead to smaller arteries, then to arterioles, then to capillaries, where oxygen, nutrients, hormones, and other substances move from the blood to the cells. It is through the capillary walls that waste products such as carbon dioxide and nitrogenous wastes are picked up. Other substances enter the blood through the capillary walls, such as hormones that are secreted by specific tissues and nutrients that are released from the intestine and liver. During one's life, due to the pumping action of the heart, blood continually circulates throughout the body maintaining the constant exchange of materials.

After passing through the capillary beds in the tissues, the blood moves into venules and on to larger veins. These eventually empty into the **anterior vena cava**, which collects all the blood from the head, neck, and arms, and the **posterior vena cava**, which drains the blood from the rest of the body. The anterior and posterior vena cavas empty into the right atrium, completing one cycle.

The portion of the circulatory system that travels from the heart to the lungs and back is known as the **pulmonary circulatory system**. The other

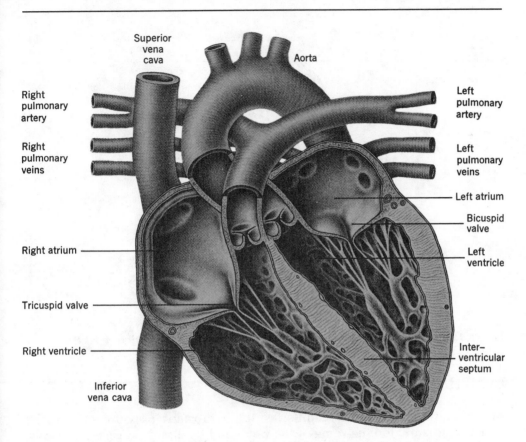

Superior
vena
cava

Aorta

Right
pulmonary
artery

Right
pulmonary
veins

Left
pulmonary
artery

Left
pulmonary
veins

Left atrium

Bicuspid
valve

Right atrium

Left
ventricle

Tricuspid valve

Right ventricle

Inter–
ventricular
septum

Inferior
vena cava

Figure 12.4 *Cross section of a human heart showing the four chambers (two atria and two ventricles). The left ventricle has the thickest, most muscular walls, and this is the chamber responsible for pumping blood through the aorta.*

portion, which carries blood from the heart to the rest of the body and back, is called the **systemic circulatory system** (see Figure 12.5).

CAPILLARIES

Unlike the walls of the larger vessels, capillaries are only one cell thick. Made of **endothelial cells**, which together compose the **endothelium**, capillaries form a tube that encloses a passageway, or **lumen**, that is about six micrometers in diameter, wide enough to allow the passage of one red blood cell at a time. While the blood moves through the capillaries, gases, hormones, and other materials diffuse both in and out of the blood and the surrounding tissues through the cytoplasm of the endothelial cells, as well as through junctions between individual endothelial cells. In addition to diffusion, active transport and **pinocytosis** (cells engulfing small amounts of liquid or very small particles) probably help ferry dissolved materials across the endothelium.

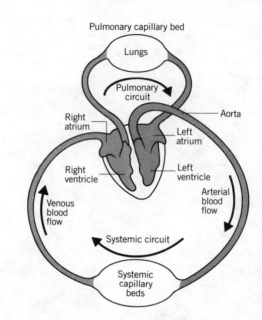

Figure 12.5 *A highly simplified representation of the human circulatory system.*

LYMPHATIC SYSTEM

At the arteriolar end of the capillaries, the hydrostatic pressure exceeds the osmotic pressure, forcing some of the water and its dissolved gases, minerals, and protein molecules through the capillary wall, or endothelium. As the blood passes along inside the capillaries, the hydrostatic pressure gradually decreases to the point where it is less than the surrounding osmotic pressure. This allows much of the surrounding fluid that was previously forced out to reenter the blood capillaries. The relatively small amount of fluid that doesn't reenter is brought back to the bloodstream by the lymphatic system.

The **lymphatic system** drains protein-containing fluids that weren't reabsorbed from the blood capillaries. These escaped fluids become the **lymph** upon being absorbed by the lymph capillaries. Lymphatic capillaries are located in the interstitial spaces between cells where the fluids accumulate. These capillaries are slightly larger and more permeable than the blood capillaries. The lymph capillaries converge into larger lymph vessels, the **lymphatics**, which resemble veins but have thinner walls and more valves. The lymphatics just underneath the skin usually run parallel to veins, while those in the viscera generally follow arteries (see Figure 12.6).

In addition to returning fluids back to the bloodstream, the lymphatic system is also an important factor in the absorption of fat from the intestines. On the other hand, blood seems to be much better at absorbing sugars and amino acids from the intestines.

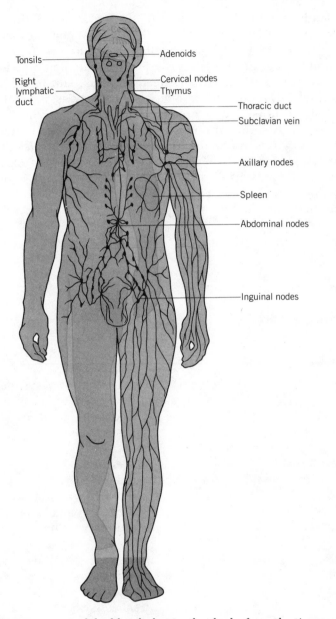

Figure 12.6 *Ninety percent of the blood plasma that leaks from the tissues passes back into the veins. The remaining 10 percent passes into the lymphatic system, where it circulates somewhat haphazardly, before being returned to the venous system.*

Located along the major lymph vessels is a series of nodes known as **lymph nodes**, or lymph glands. These are composed of a matrix of connective tissue harboring phagocytotic cells that filter out and cleanse the lymph by ingesting bacteria, cell fragments, and entire cells. The lymph glands then return these wastes to the blood, where they are carried to the lungs, kidneys, and sweat glands that eliminate them from the body. In addition, these wastes are detoxified when they pass through the liver. When the body is fighting infection, the lymph nodes become swollen and more sensitive.

In addition to filtering and processing lymph, the lymph nodes, along with other lymphoid tissues such as the **spleen, thymus,** and **tonsils,** are the sites where certain **white blood cells,** known as **lymphocytes,** are formed.

The lymph is moved by a process similar to that which moves blood through the veins. A series of one-way valves work in conjunction with the pressure applied by the contraction of nearby skeletal muscles that constantly squeeze the lymph forward in a one-way direction. Some animals, though no mammals, have specialized pumping devices, or lymphatic hearts, located along the major lymphatic vessels to move the lymph along. Ultimately the lymph is emptied into the left and right subclavian veins, both located in the base of the neck.

Key Terms

alveoli
anterior vena cava
aorta
aortic semilunar valve
arteries
arterioles
atrium
bicuspid valve
capillaries
capillary beds
cardiovascular system
closed circulatory system
deoxygenated blood
dorsal longitudinal vessel
encapsulation
endothelial cells
endothelium
haemocoel
haemocytes
haemolymph
heart chambers
hearts
heart valves
hemocyanin
interstitial fluids
lumen
lungs

lymph
lymphatics
lymphatic system
lymph nodes
lymphocytes
Malpighian tubules
mitral valve
open circulatory system
ostia
oxygenated blood
phagocytosis
pinocytosis
posterior vena cava
pulmonary artery
pulmonary circulatory system
pulmonary semilunar valve
pulmonary veins
spleen
systematic circulatory system
thymus
tonsils
tracheal system
tricuspid valve
veins
ventricle
venules
white blood cells

Chapter 12 Self-Test

QUESTIONS TO THINK ABOUT

1. Describe the similarities and differences between closed and open circulatory systems.
2. How does the insect tracheal system affect its circulatory system?
3. What is the role of the Malpighian tubules, and how do they interact with the circulatory system of what kinds of animals?
4. Compare and contrast several different types of hearts.
5. Follow the entire route of blood through a vertebrate circulatory system.
6. What is the function of a lymphatic system?

MULTIPLE-CHOICE QUESTIONS

Higher Circulation

1. Animals with open circulatory systems include the following:

   a. spiders and insects
   b. crabs and lobsters
   c. snails and clams
   d. a and b
   e. a, b, and c

2. Animals with closed circulatory systems include the following:

   a. annelids
   b. fish, amphibians, and reptiles
   c. birds and mammals
   d. b and c
   e. a, b, and c

3. Unlike most organisms, insects have a network of hollow tubes to carry air through pores perforating the body wall to the cells throughout the insect's interior; these tubes are called _____ .

   a. exoskeleton
   b. tracheids
   c. tracheae
   d. ostia
   e. osteids

4. Insects have blood that circulates through an open internal body cavity known as the _____ .

   a. haemolymph
   b. pericardium
   c. endocardium
   d. endometrium
   e. haemocoel

5. Insect blood, the _____ , is composed of water, plasma, minerals, nutrients, waste products, and blood cells, or _____ .

   a. lymph, nematocysts
   b. lymphocytes, erythrocytes
   c. haemolymph, haemocytes
   d. thromboplasma, thrombocytes
   e. lymph, erythrocytes

6. In insects, blood enters a _____ , which is often called the heart, through openings, or _____ .

   a. ventral tube, pores
   b. ventral longitudinal vessel, lenticels
   c. dorsal longitudinal vessel, ostia
   d. atrium, ducts
   e. ventricle, Malpighian tubules

7. Waste products enter an insect's alimentary canal through blind sacs known as _____ .

   a. ostia
   b. Malpighian tubules
   c. pores
   d. lenticels
   e. dorsal longitudinal vessels

8. The blood of many arthropods contains an oxygen-carrying pigment, _____ , that is like that found in vertebrate blood.

   a. hemoglobin
   b. hemocyanin
   c. gammaglobulin
   d. albumin
   e. anthocyanin

Circulation in Vertebrates

9. Fish hearts have _____ internal chambers arranged linearly.

   a. one
   b. two
   c. three
   d. four
   e. six

10. Amphibians have a _____ chambered heart, with _____ atria and _____ ventricles.

   a. two, one, one
   b. three, one, two
   c. three, two, one
   d. four, two, two
   e. four, one, three

11. Reptiles have _____ atrium(a) and _____ ventricle(s).

    a. one, one
    b. two, one
    c. one, two
    d. two, two
    e. two, three

12. The ventricular division is complete in the following hearts:

    a. birds
    b. mammals
    c. reptiles
    d. amphibians
    e. a and b

13. In the human heart, when the blood is forced from the right atrium to the right ventricle, it moves through the _____ .

    a. tricuspid valve
    b. pulmonary semilunar valve
    c. bicuspid valve
    d. mitral valve
    e. aortic semilunar valve

14. When the blood moves from the right ventricle into the pulmonary artery, it passes through the _____ .

    a. tricuspid valve
    b. pulmonary semilunar valve
    c. bicuspid valve
    d. mitral valve
    e. aortic semilunar valve

15. The left ventricle contracts and pumps blood through another valve, called the _____ , into the artery called the _____ .

    a. mitral valve, right vena cava
    b. bicuspid valve, inferior vena cava
    c. mitral valve, superior vena cava
    d. aortic semilunar valve, aorta
    e. aortic semilunar valve, inferior vena cava

16. After passing through the capillary beds in the tissues, the blood moves into venules and on to larger veins, which eventually empty into the _____ .

    a. anterior vena cava
    b. superior vena cava
    c. aorta
    d. a or b
    e. a, b, or c

17. The portion of the circulatory system travelling from the heart to the lungs and back is known as the _____ , and the other portion of the circulatory system carrying blood from the heart to the rest of the body and back is called the _____ .

    a. systemic circulation, pulmonary circulation
    b. pulmonary circulation, systemic circulation
    c. cardiac circulation, somatic circulation
    d. cardio-pulmonary circulation, somatic circulation
    e. cardio-pulmonary circulation, cardio-systemic circulation

Capillaries and Lymphatics

18. In humans and other vertebrates, the blood is enclosed within a system composed of a heart and vessels, where the heart pumps the blood into large _____ that branch into smaller _____ that branch into the smallest blood vessels, the _____ .

    a. arteries, arterioles, capillaries
    b. veins, venules, capillaries
    c. capillaries, arterioles, arteries
    d. capillaries, venules, veins
    e. a and c

19. Capillary walls are the following number of cells thick:

    a. one
    b. two
    c. three
    d. four
    e. five

20. The walls of the capillaries are made of _____ .

    a. endothelial cells
    b. endometrium cells
    c. endomysium cells
    d. endoneurium cells
    e. endocrine cells

21. Escaped fluids from the blood capillaries become the _____ upon being absorbed by the _____ .

    a. lymph, lymph capillaries
    b. blood, blood capillaries
    c. plasma, plasma capillaries
    d. plasma, lymphatics
    e. plasma, lymph nodes

22. Located along the major lymph vessels is a series of nodes known as
   _____ .

   a. plasma nodes
   b. lymph nodes
   c. lymph glands
   d. a and b
   e. b and c

23. Lymph glands do the following:

   a. filter out and cleanse the lymph
   b. harbor phagocytotic cells
   c. contain cells that ingest bacteria
   d. contain cells that ingest cell fragments and dead cells
   e. all of the above

ANSWERS

1. e	7. b	13. a	19. a
2. e	8. b	14. b	20. a
3. c	9. d	15. d	21. a
4. e	10. c	16. d	22. e
5. c	11. d	17. b	23. e
6. c	12. e	18. a	

CHAPTER 13

Blood

Like muscle, bone, cartilage, and nerves, **blood** is a type of tissue, a collection of similar types of cells and the associated intercellular substances that surround them. Tissues are categorized into four main types: 1) epithelium, 2) muscle, 3) nervous, and 4) connective. Connective tissue includes blood, lymph, bone, and cartilage. In the case of blood and lymph, the base substance, or **matrix**, is a liquid. What differentiates these two from other tissue types is that they are not stationary. Blood is a fluid flowing through blood vessels throughout the body (lymph runs through the lymph vessels).

Blood accounts for about 8 percent of a human's total body weight, amounting to an average of four to six liters per adult (over a gallon), depending on individual size. Blood is thicker (more viscous) and slightly heavier than water. And, depending on the organism, blood is usually slightly warmer than the animal's body temperature. While the core body temperature of most humans is 37°C (98.6°F), their blood is about 38°C (100.4°F). Blood pH is slightly alkaline, ranging from about 7.35 to 7.45. Its salt (NaCl) concentration normally varies from about 85 to 90 ppt (parts per thousand), or two to three times the concentration of sea water, which is usually about 34 ppt.

Plasma is the fluid portion of both blood and lymph. Fifty to sixty percent of the blood volume consists of plasma. When the proteins involved in clotting are removed from blood plasma, the remaining liquid is called **serum**. The plasma, which is over 90 percent water, carries a variety of ions and molecules. In addition to salts and proteins, there are many nutrients such as amino acids, fats, and glucose. There are also dissolved gases such as carbon dioxide, as well as antibodies, hormones, enzymes, and certain waste products such as urea and uric acid. The relative amount of plasma in the blood depends upon the species, the sex, the organism's health when being examined, and on a host of other variables. The remaining 40 to 50 percent of the blood volume is composed of cells and cell fragments that can be divided into three main categories: **red blood cells**, or **erythrocytes**; **white blood cells**, or **leukocytes**; and **platelets**, or **thrombocytes**, which are fragments of cells.

ERYTHROCYTES (RED BLOOD CELLS)

The cells in human blood are often referred to as corpuscles. Of these, 99 percent are red blood cells, which are involved in the transportation of oxygen throughout the body. Quite distinct in appearance, they are biconcave, flattened disks (see Figure 13.1). Birds and reptiles have red blood cells that retain their nucleus throughout the lifespan of each red blood cell, but when mammalian erythrocytes mature, their nuclei disintegrate.

It has been surmised that the loss of the nuclei adds to the cells' oxygen exchange capacity. The biconcave design of red blood cells increases the surface area so that **hemoglobin**, the iron-containing protein that functions in oxygen transport, can fill almost the entire volume of the mammalian red blood cells, enabling them to interact more effectively in a deoxygenated medium. Having an average lifespan of about three to four months, red blood cells are then consumed by the liver. New red blood cells are constantly made by the red bone marrow.

The concentration of cells and particulate matter in the blood plasma, known as the **hematocrit value**, is usually about 40 to 45 percent. If the concentration falls below this level, the individual is in danger of not having enough red blood cells or hemoglobin to supply the proper amount of oxygen to maintain cellular respiration. When the blood's concentration of red blood cells, or the red blood cell's concentration of hemoglobin, drops below normal, a person becomes **anemic**; the condition is known as **anemia**. Symptoms often include a general pallor (pale appearance) and lack of energy.

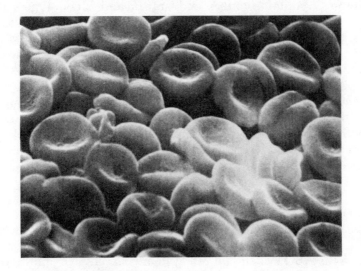

Figure 13.1 A mass of human red blood cells. Their flattened, indented shape enhances their flexibility and surface area/volume ratio, improving the efficiency with which they carry and exchange oxygen and carbon dioxide.

BLOOD TYPES

Everyone has highly individualized blood that is directly attributable to proteins and other genetically determined factors located both on the surfaces of red blood cells and in the plasma bathing the red blood cells. Of all the blood groups (there are over 300), the most widely known is the ABO group. When doctors first began giving blood transfusions (transferring blood from one person to another), it became apparent that while many patients' lives were saved, many other patients died almost immediately. It took years until it was learned that there were different blood types, some of which are incompatible.

When incompatible blood types are mixed, the blood of the patient receiving the transfusion clumps and clots, blocking capillaries, clogging organs, and causing strokes. We now understand that some erythrocyte surfaces contain genetically determined types of proteins known as the **agglutinogenic proteins**, which function as **antigens**, substances that stimulate the body to produce antibodies against it. **Antibodies** are proteins that inactivate or destroy antigens. Antigens, in the form of proteins on the red blood cell surface, and plasma antibodies both account for different blood types.

The ABO group depends on the combination of two alleles at a chromosomal locus, which can be AA, AB, BB, AO, BO, or OO. These are the result of two different antigens, A and B, and their absence, which is called O. Two AA or one A without another antigen (AO) produce type A blood. BB and BO produce type B blood. When A and B are found together, the blood is called type AB, and when neither occurs, the blood is type O.

Blood type A contains antibody B (anti-B); blood types B contains anti-A, blood type O contains anti-A and anti-B; and blood type AB contains neither anti-A nor anti-B. When antibodies are present in the plasma of one blood type, they will react with the antigens on the erythrocytes of another blood type, resulting in clumping. Therefore it is always best, when getting a blood transfusion, to obtain the same blood type of blood as your own. Because type O can be given to people with any blood type, it is often called the **universal donor**. Unless a transfusion is going to be massive, usually types A or B can be given to a person with type AB. Likewise, people with types A or B can receive blood from a person with AB. Table 13.1 shows the antigens and antibodies found in each blood type, and which types of blood can act as a donor during a blood transfusion.

RH

The Rh system derives its name from rhesus monkeys among which it was first studied. Like the ABO system, the **Rh system** is based on agglutinogens on the erythrocyte surface. People with erythrocytes having the Rh agglutinogens are called Rh$^+$; those without it are Rh$^-$. When blood is transfused, both the ABO

Table 13.1 *Compatiblity of Blood Types*

Antigens and Antibodies

Blood Type	Antigens on the Erythrocytes	Antibodies in the Plasma
A	A	anti-B
B	B	anti-A
AB	A and B	None
O	None	anti-A and anti-B

Donor/Receiver Relationships

Blood Type	Can Receive Blood from	Can Act as Donor to
A	A, O	A, AB
B	B, O	B, AB
AB	A, B, AB, O	AB
O	O	A, B, AB, O

blood types and the Rh type must be checked, because Rh incompatibility can produce a severe reaction that sometimes is fatal.

Rh types are also important in regard to having children. When the father is Rh$^+$, the mother is Rh$^-$, and the fetus is Rh$^+$, fetal Rh$^+$ antigens may enter the mother's blood during delivery. The fetal Rh$^+$ antigens will stimulate the mother's blood to make anti-Rh antibodies. Then, should the woman become pregnant again, her anti-Rh antibodies will cross the placenta, resulting in **hemolysis**, or the breakage of fetal erythrocytes, which will release the hemoglobin. This condition, known as **erythroblastosis fetalis**, used to kill the baby. Later, when a baby was born with such a condition, the doctors slowly removed its blood, replacing it with antibody-free blood. Today, erythroblastosis is routinely prevented by injecting an anti-Rh preparation into Rh$^-$ mothers right after delivery or an abortion. This injection contains antibodies that rid the mother's blood of the fetal agglutinogens. This prevents the mother's blood from producing its own antibodies and, thus, protects a fetus during the next pregnancy.

WHITE BLOOD CELLS (LEUKOCYTES)

In each organism the mature red blood cells are all basically identical to one another. This is not true of an animal's mature white blood cells (also called leukocytes). These are separated into several distinct categories. There are **basophils, eosinophils, neutrophils, lymphocytes,** and **monocytes** (see Figure 13.2). All have nuclei and none contain hemoglobin. Of the five types listed above, the first three develop in the red bone marrow. The other two develop from lymphoid and myeloid tissues.

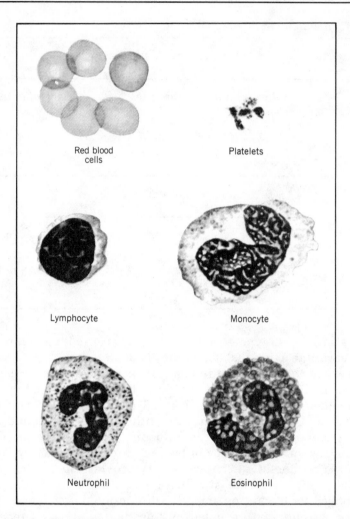

Figure 13.2 *Blood cell types are classified according to the shape and size of the cell; absence, presence, shape, and size of the nucleus; and whether the cytoplasm is granular. These blood cell differences reflect their division of labor.*

Generally, leukocytes combat inflammation and infection, either by ingesting bacteria and dead matter through phagocytosis, or by producing chemicals that inactivate foreign bodies. Important factors in the **immune system**, lymphocytes are involved in producing antibodies to inactivate antigens. Most antigens are not produced by the body but are released by bacteria or other foreign microorganisms. However, when blood is transfused from one person to another, specific proteins in an erythrocyte's surface can be an antigen.

There are two basic types of lymphocytes. These are differentiated according to their function. Although they originate in the bone marrow, some lymphocytes, while developing, pass through the thymus gland. Therefore, they are called **T cells**. They specialize in combating foreign cells or tissues introduced into the body such as bacteria, fungi, and cells in transplanted organs. They are also involved in fighting cancer cells and in protecting cells from viruses. T cells are involved both in forming cell-killing chemicals and in attracting other white blood cells that engulf the foreign bodies encountered. They do not release antibodies.

The other lymphocytes, those that don't pass through the thymus gland during their developmental stage, are called **B cells**. When stimulated, they divide and form plasma cells that manufacture antibodies. When the antibodies encounter antigens, they attempt to destroy them. This is known as the **antigen-antibody reaction**. In addition to functioning inside the bloodstream, most leucocytes possess the ability to pass through minute spaces in the capillary walls and nearby tissues.

BLOOD TEST

A **blood test** often involves a **differential count**, which amounts to counting the number of each kind of white blood cell. A normal differential count contains 60 to 70 percent neutrophils, with considerably fewer of the other types of white blood cells. White blood cell counts that vary significantly from the norm often indicate the presence of an injury or infection. Such information is valuable to a doctor attempting to diagnose the cause or nature of a disorder from the symptoms.

Because the leukocytes are constantly involved in fighting unhealthy cells, such as those which may be cancerous, as well as bacteria and other foreign substances that enter the body through such points as the mouth, nose, and skin, leukocytes usually have very short life spans. Their normal life span is considerably shorter than that of the normal red blood cell, generally lasting only a few days. During a period of infection, they may survive only a few hours.

PLATELETS (THROMBOCYTES)

Platelets are small, disc-shaped, anucleate (without a nucleus) cells, or cell fragments, that are crucial to the clotting process. They help minimize the amount of bleeding when a blood vessel is cut or damaged by initiating the steps that result in coagulation. The damaged tissues and ruptured platelets both release substances known as **thromboplastins**. These are thought to initiate blood clotting in the presence of calcium ions.

Once released, the thromboplastins convert the plasma protein prothrombin into **thrombin**, which then converts **fibrinogen**, another plasma protein, into fibrin. Fibrin is fibrous and forms a meshwork that begins to shrink, squeezing out fluids to form a blood clot. These steps are summarized on the following page.

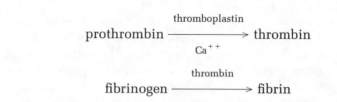

1.
$$\text{prothrombin} \xrightarrow[\text{Ca}^{++}]{\text{thromboplastin}} \text{thrombin}$$

2.
$$\text{fibrinogen} \xrightarrow{\text{thrombin}} \text{fibrin}$$

Recently, the blood clotting process has been explained more fully. It has been shown that when blood comes in contact with cells outside the bloodstream, a protein molecule called **tissue factor** (TF), which is found on the surface of most cells outside the bloodstream, initiates the sequence of events leading to coagulation.

This tissue factor protein combines with a protein that circulates in the blood (factor 7). Instantly this becomes a new tissue factor, 7a (TF/7a), which initiates the clotting process by binding with tissue factor 10 and converting it into factor 10a. This binds to protein factor 5a. Then, these bind to the circulation protein prothrombin, creating thrombin molecules, which, in turn, create fibrin molecules, as explained above.

To aid the clotting process, damaged tissues release **histamine**, a chemical that expands the diameter of the capillaries and arterioles, thereby increasing the blood supply to the area and permitting more clotting proteins to leak out of the bloodstream into the damaged tissue. In addition to stopping the bleeding, clotting blocks bacteria and other foreign agents from entering a wound. The dead white blood cells that accumulate around the inflamed wound constitute the thick, creamy yellow fluid known as **pus**. If the wound is near the surface, the damaged area may eventually open, allowing the pus to drain. Otherwise the body slowly resorbs the dead cells.

IONS, SALTS, AND PROTEINS IN THE BLOOD

The inorganic ions and salts comprise approximately 0.9 percent of a human's blood plasma by weight. More than two-thirds of this is sodium chloride (NaCl, ordinary table salt). The most abundant positively charged inorganic ions (cations) found in the plasma are sodium (Na^+), calcium (Ca^{2+}), potassium (K^+), and magnesium (Mg^{2+}). The primary negatively charged inorganic ions are chloride (Cl^-), bicarbonate (HCO_3^-), phosphate (HPO_4^{2-} and $H_2PO_4^-$) and sulfate (SO_4^{2-}).

The plasma proteins contribute another 8 to 9 percent of the plasma by weight. Most of these, such as the albumins, globulins, and fibrinogen, are synthesized by the liver. They are important in maintaining the blood's osmotic pressure, which is particularly important in the capillary beds. They are also crucial for maintaining the viscosity of the plasma, which is vital to the heart for maintaining a normal blood pressure.

HEARTBEAT

The heartbeat is initiated in its unique tissue, the **nodal tissue**, which has characteristics similar to both muscles and nerves, being able to contract like muscles and transmit impulses like nerves. There are two **nodes**, or areas composed of nodal tissue, in the heart. The heartbeat begins in the **sino-atrial node** (or S-A node), also called the **pacemaker**, which is located in the wall of the right atrium near the anterior vena cava. At regular intervals, a wave of contraction spreads from the S-A node across the walls of the atria to the **atrio-ventricular node** (A-V node), located in the lower part of the partition between both atria. As the wave of contraction reaches the A-V node, this node transmits impulses to the ventricles via a bundle of branching fibers known as the **bundle of His**. The impulse is slowed down in the bundle of His. During atrial relaxation, ventricular contraction occurs (see Figure 13.3).

These contractions occur at regular intervals, which are modified according to the physiology of the organism at the time. The speed at which these contractions occur is measured by counting the number of cycles per minute; this is known as the **pulse rate**, which in adult humans averages 70 beats per minute, though it varies considerably among individuals.

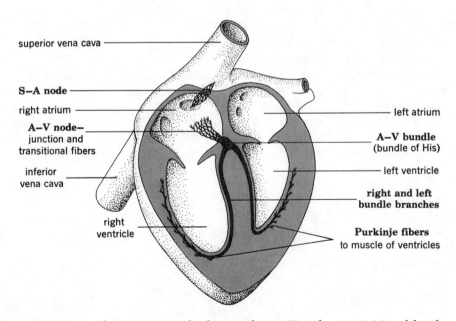

Figure 13.3 *Pacemaker tissues in the human heart. Heartbeat is initiated by the S-A node, which stimulates the atria to contract. The A-V node coordinates these contractions with those of the ventricles.*

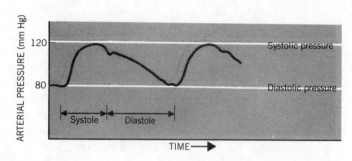

Figure 13.4 An EKG (electrocardiogram) tracing a heartbeat, measuring the systolic and diastolic pressure.

BLOOD PRESSURE

Blood pressure is defined as the pressure exerted by blood on the wall of any blood vessel. Two pressures are measured, that during the **systole**, when the ventricles are contracting, and that during the **diastole**, when the ventricles are relaxed (see Figure 13.4). The pressure is measured with a **sphygmomanometer**, a device that transfers the blood pressure into pressure that pushes mercury (Hg) up a tube. The average systolic blood pressure of a young adult male is about 120 mm of mercury; the average diastolic pressure is 80 mm of mercury. This is usually represented as 120/80. Young adult females have blood pressures that usually average about 10 mm of mercury less for both the systolic and diastolic pressures.

Blood pressure tends to increase with age, but it also increases because of certain inherited and dietary variables. Because accumulated evidence shows a correlation between high blood pressure and increased risk of heart attack, many doctors routinely take their patient's blood pressure to monitor any potential risk. Advice about diet and exercise can be given to patients who fall into the risk category. Certain medications have also been shown to be helpful.

Key Terms

agglutinogenic protein	bundle of His
anemia	diastole
anemic	differential count
antibodies	eosinophils
antigen-antibody reaction	erythroblastosis fetalis
antigens	erythrocytes
atrio-ventricular node	fibrin
A-V node	fibrinogen
basophils	hematocrit value
B cells	hemoglobin
blood	hemolysis
blood pressure	histamine
blood test	immune system
blood types	leukocytes

lymphocytes
matrix
monocytes
neutrophils
nodal tissue
nodes
pacemaker
plasma
platelets
prothrombin
pulse rate
pus
red blood cells
Rh system

S-A node
serum
sino-atrial node
sphygmomanometer
systole
T cells
thrombin
thrombocytes
thromboplastin
tissue
tissue factor
universal donor
white blood cells

Chapter 13 Self-Test

QUESTIONS TO THINK ABOUT

1. Analyze, compare, and contrast the contents of the blood of a vertebrate and an insect.
2. How does blood clot?
3. How does a heart beat?
4. What is blood pressure?

MULTIPLE-CHOICE QUESTIONS

1. The fluid flowing through all the vessels of the body, excluding the lymph vessels, is the _____ .

   a. serum
   b. lymph
   c. blood
   d. plasma
   e. lumen

2. Connective tissue includes the following types of tissues:

   a. nervous and connective
   b. epithelium and muscle
   c. muscle and nervous
   d. bone and epithelium
   e. lymph and cartilage

3. _____ is the fluid portion of both blood and lymph.

   a. plasma
   b. matrix
   c. ground substance
   d. water
   e. urea

4. _____ is the fluid portion of both blood and lymph.

   a. plasma
   b. matrix
   c. ground substance
   d. water
   e. urea

5. When the proteins involved in clotting are removed from blood plasma, the remaining liquid is called _____ .

   a. water
   b. urea
   c. ground substance
   d. serum
   e. lumen

6. In addition to the plasma, the remaining 40 to 50 percent of the blood volume is composed of the following:

   a. thrombocytes and platelets
   b. erythrocytes and leukocytes
   c. white blood cells and red blood cells
   d. leukocytes
   e. all of the above

Erythrocytes, Leukocytes, and Blood Types

7. The concentration of cells and particulate matter in the blood plasma is known as the _____ .

   a. erythrocyte count
   b. corpuscle count
   c. hemoglobin level
   d. anemia analysis
   e. hematocrit value

8. When the blood's concentration of red blood cells or the red blood cell's concentration of hemoglobin drops below normal, a person becomes _____ .

   a. anemic
   b. weak
   c. pale
   d. all of the above
   e. none of the above

9. The surfaces of erythrocytes contain genetically determined types of proteins that are responsible for the two major blood group classifications,
   which are:

   a. ABO and Rh
   b. Type A and Type B
   c. Type AB and O
   d. Rh^- and Rh^+
   e. Type O and Rh^-

10. Blood Type O will not agglutinate with any of the other blood types and
    therefore is known as the ⎯⎯⎯⎯⎯⎯ .

    a. universal donor
    b. Blood Type A
    c. Blood Type B
    d. Blood Type AB
    e. Blood Type ABO

11. Lymphocytes are involved in producing certain proteins known
    as ⎯⎯⎯⎯⎯⎯ that inactivate foreign chemicals known as
    ⎯⎯⎯⎯⎯⎯ .

    a. antigens, antibodies
    b. antibodies, antigens
    c. basophils, eosinophils
    d. neutrophils, monocytes
    e. none of the above

Platelets, Ions, Salts, and Proteins in the Blood

12. Both cut or damaged tissues as well as ruptured platelets release sub-
    stances involved in clotting known as ⎯⎯⎯⎯⎯⎯ .

    a. thrombins
    b. prothrombins
    c. fibrinogens
    d. calcium ions
    e. thromboplastins

13. The presence of ⎯⎯⎯⎯⎯⎯ ions is needed for thromboplastins to
    initiate the blood clotting process.

    a. magnesium
    b. cadmium
    c. lead
    d. calcium
    e. chloride

14. Once released, the thromboplastins convert the plasma protein _____ _____ into _____.
    a. fibrin, fibrinogen
    b. prothrombin, thrombin
    c. thrombinogen, thrombin
    d. thrombin, prothrombin
    e. profibrinogen, fibrinogen

15. Thrombin converts another plasma protein into _____.
    a. fibrin
    b. histamine
    c. calcium ions
    d. prothrombin
    e. thromboplastin

16. To aid the clotting process, damaged tissues release _____, a chemical that expands the diameter of the capillaries and arterioles, thereby increasing the blood supply to the area, permitting more clotting proteins to leak out of the bloodstream into the damaged tissue.
    a. histamine
    b. fibrin
    c. calcium ions
    d. thrombin
    e. thromboplastin

17. The most abundant positively charged inorganic ions found in the plasma include:
    a. sodium
    b. calcium
    c. potassium
    d. magnesium
    e. all of the above

18. The primary negatively charged inorganic ions found in the human blood plasma include the following:
    a. chloride
    b. bicarbonate
    c. phosphate
    d. sulfate
    e. all of the above

Heartbeat

19. The heartbeat begins in the _____.
    a. sinoatrial node
    b. S-A node
    c. pacemaker
    d. all of the above
    e. none of the above

20. A wave of contraction spreads from the S-A node across the walls of the atria to the _____ which transmits impulses to the ventricles via _____ .

    a. atrio-ventricular node, a bundle of branching fibers
    b. A-V node, bundle of His
    c. atrio-ventricular node, bundle of His
    d. A-V node, a bundle of branching fibers
    e. all of the above

21. The blood pressure exerted on the wall of any blood vessel when the ventricle is contracted is the _____ , and the pressure exerted on the wall of any vessel when the heart is relaxed is the _____ .

    a. systolic pressure, diastolic pressure
    b. diastolic pressure, systolic pressure
    c. sphygmomanomic pressure, blood pressure
    d. ventricular pressure, aventricular pressure
    e. ventricular pressure, auricular pressure

ANSWERS

1. c	7. e	13. d	19. d
2. e	8. d	14. b	20. e
3. a	9. a	15. b	21. a
4. a	10. a	16. a	
5. d	11. b	17. e	
6. e	12. e	18. e	

CHAPTER 14

Nutrition

Catering to a huge, lucrative market, popularizations of health and nutrition have been responsible for a constant flow of misinformation. One should therefore take much that is being said with a grain of salt. With diet and nutrition books on the best-sellers' lists, some publishers appear less concerned with what's right than with what sells.

To understand nutrition, you have to understand that it is a combination of the many integrated processes involved in an organism's absorption and use of materials. Most of the actual absorption takes place on a molecular level. Organisms must procure the food, which is then broken down to chemical constituents that are absorbed by the cells. These atoms and molecules provide the raw materials that constitute, maintain, and run the organisms. The energy produced by the raw materials is usually measured in units known as **calories**; one calorie is the amount of energy needed to raise one gram or cubic centimeter (cc) of water one degree centigrade. The calories usually referred to in biology and nutrition are kilogram-calories (Kcal); each is 1000 gram-calories. When calories are discussed below, they are kilogram-calories; that is, the amount of heat required to raise the temperature of 1000 grams (one kilogram) of water from 14.5°C to 15.5°C.

Breaking down the nutrient materials we call food involves a series of processes that in the aggregate are known as **digestion**. Digestion is the physical and/or chemical breakdown of food into sizes small enough to be passed across cell membranes. Our food is the result of other organisms' procurement of nutrients from the environment. Eating other organisms, or their parts, saves us considerable time and energy.

Green plants and a number of additional life forms can survive without consuming other organisms. Animals, however, could not exist without the green plants that synthesize the majority of molecules that are passed on to other organisms. (In addition, it should be remembered that green plants generate oxygen and consume carbon dioxide, while most animals consume oxygen and generate carbon dioxide.)

Of the four basic macromolecules found in living things, **proteins, carbohydrates**, and **fats** are so abundant, readily accessible, and nutritionally useful that they have become the primary molecular units representing the food in our diets. **Nucleic acids**, the fourth group of macromolecules, are not abundant enough to be an important dietary item. Of course, water and many minerals are also important and will be discussed in this chapter.

PROTEINS

Proteins are one of the most plentiful organic substances in animals. Some are among the largest molecules that exist naturally in living things; some are also quite small. While they are widely used as structural materials to form muscles and organs, they also form hair, fingernails, feathers, and scales.

The protein we eat has to be digested to produce more readily absorbable chemical constituents. Proteins are composed of subunits, **amino acids**, arranged in a pattern specific to each distinct protein type. The cells in human tissues are able to synthesize only 10 of the 20 amino acids required in our proteins. The other 10 are obtained from plant and animal tissues present in the diet; these are known as the **essential amino acids**.

When arranged in their specific patterns, the amino acids make thousands of different kinds of proteins. All amino acid molecules contain one **amino group** ($-NH_2$). They also contain a **carboxyl group** ($-COOH$), which is a characteristic of organic acids. Each of these groups is connected to a carbon atom. Two amino acids are held together by a specific type of **covalent bond** (resulting from a shared pair of electrons) called a **peptide bond**. Peptide bonds are formed by joining the amino group of one amino acid with the carboxyl group of the other, through the elimination of water. This is known as a **dehydration**, or **condensation**, **reaction**.

Along with removing excess simple sugars and processing them into glycogen, which is discussed below in the carbohydrate section, organisms also process fatty acids and other lipids, as well as proteins. Animals store amino acids, proteins, and other nitrogenous compounds that can be used for energy when carbohydrates and fat supplies are exhausted. The first step in converting the nitrogenous compounds into glucose involves removing an amino group ($-NH_2$), a step called **deamination**. During deamination, the amino group is converted into ammonia (NH_3), a waste product that is converted into a less toxic form and then excreted.

CARBON

Molecules based on carbon are called **organic molecules**. The properties of carbon make it one of the most important elements, upon which almost all molecules composing living organisms are based. This element is extremely versatile since a seemingly endless variety of molecules can be formed around it. Four atoms may bond to each carbon atom at any given time; carbon atoms will bond with many other elements. Carbon atoms can join to one another to form long chains. These chains may have additional carbon chains branching from them. Occasionally the ends of the carbon chains join, forming rings. It seems that carbon is the only atom that forms enough different, complex, stable compounds to be the base for the variety of molecules found in living organisms.

FUNCTIONAL GROUPS

Several different **functional groups**, such as the amino group and the carboxyl group mentioned above, are used to categorize the wide range of carbon-based molecules into their molecular "families." Each functional group causes the molecule to behave in a specific manner. Members of each molecular family predictably behave in certain ways, taking part in chemical reactions no matter what the rest of the molecule is like. For instance, when a molecule has the carboxyl group, it is called an acid because it is a proton donor, releasing hydrogen ions when in water. Some of the groups that identify important organic molecular families are presented in Figure 14.1.

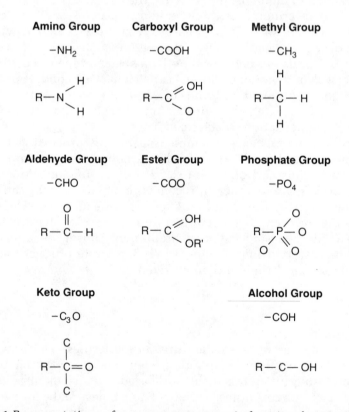

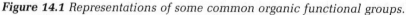

Figure 14.1 *Representations of some common organic functional groups.*

MONOMERS AND POLYMERS

All the molecules without any carbon are termed **inorganic**. Of the **organic** molecules, those with carbon, the smallest molecular units are **monomers**. When these monomer units are linked together, the molecules are called **polymers**. The polymers are sometimes called macromolecules. They can be long, straight, or branched chains. A myriad of different polymers, or **polypeptides** when speaking of proteins, can be formed from the relatively few monomer units that exist. Some proteins consist of just one polypeptide chain. Others are made from two or more polypeptides linked together.

These macromolecular, polymeric compounds, which compose the bulk of organic matter in animal cells, are important constituents of cell membranes and connective tissues between tissues and organs. Proteins are important to both plants and animals because they are vital to the plasma membranes and intracellular organelles. During certain stages in some plants' life histories, such as in the seeds of legumes (the plants in the bean and pea family), the protein concentrations reach very high levels.

In addition to their importance in the structural components of animals, there are many proteins that have other functions. For instance, there are a number of types of proteins that are important in blood. Some help in transporting calcium ions, and others are vital in the clotting process. Others transport amino acids and various chemicals to parts of the body requiring them at any given time. Antibodies are largely made of proteins.

Enzymes, another group of proteins, have been shown to have specific shapes that enable them to affect all the chemical reactions that occur inside cells. They accomplish this by attaching themselves to specific molecules to bring them together in an orderly manner so that the chemical reactions can proceed step by step. The enzymes that bring molecules together, as well as break them apart, are **catalysts**. Specifically, enzymes are catalysts made of protein. Many of the chemical reactions that break down food molecules require specific enzymes.

The body doesn't need much protein each day. The Recommended Dietary Allowance (RDA), published by the National Academy of Sciences in conjunction with the National Research Council, states that an adult needs about 60 grams of protein per day, which is about two ounces. A one-year-old needs about 25 grams; otherwise symptoms of protein deficiency may occur. Depending on the degree of the symptoms, a protein deficiency may be diagnosed as **kwashiorkor**. Adults with this disease slowly lose their muscle tissue. The process can be reversed by adding protein to their diet. Children suffering from kwashiorkor may recover if treated in time, but they may never grow as tall as they might have otherwise, and their learning abilities may be diminished.

CARBOHYDRATES

Carbohydrates form another class of organic compounds. They are composed of carbon, hydrogen, and oxygen, usually with the ratio of two hydrogen atoms to one oxygen atom to one carbon atom. Carbohydrates are rich in energy, but organisms cannot survive long on just a carbohydrate diet. Most diets consisting of just carbohydrates and fats would eventually prove fatal because neither of these nutritional groups contain the nitrogen necessary to create the amino acids that form the proteins. Unlike most animals, plants have the capacity to combine inorganic nitrogen, usually in the form of nitrate, with carbon from the carbohydrates they manufacture through photosynthesis. (Photosynthesis is the process by which plants use light energy to convert water and carbon dioxide into carbohydrates.) From these, plants make their own proteins.

SIMPLE SUGARS

Carbohydrates all have a very similar molecular arrangement, though they range in size from the smallest simple sugars, **monosaccharides**, to the larger **polysaccharides**, which form such molecules as starch and cellulose. The length of the carbon chain in simple sugars varies, depending on the specific sugar involved. Some contain just three carbons: Those containing three carbons are the **trioses**, and those with four carbons are **tetroses**. The trioses, **pentoses**, and **hexoses** are biologically important because they are the monomers most often involved in the complex, polymeric carbohydrates.

DIGESTION

Food is chewed in the mouth, where it is mixed with **saliva**, which contains **amylase**, a starch-digesting enzyme. Once chewed, food is swallowed, during which it passes through the throat to the **stomach**, via the **esophagus**, by a wave of muscle contractions known as **peristalsis**. The **stomach**, located in the upper portion of the abdominal cavity, stores food, making discontinuous feeding possible. Its muscular lining also creates an environment where chewing, mixing, and further mechanical breakdown contributes to the foods' digestion.

The stomach secretes mucous to protect itself from the mixture of hydrochloric acid and digestive enzymes, the **gastric juice**, which it also secretes. Once the stomach's contents are reduced to a soupy mixture, they pass through the narrow muscular passageway, the **pyloric sphincter**, into the first portion of the **small intestine**, known as the **duodenum**. Digestive enzymes from the **pancreas** and the **gall bladder** are released into the duodenum, from which the contents move on through the rest of the small intestine, where most of the remaining digestion and nutrient absorption occurs.

Much of the nutritional contents of the small intestine are absorbed into the bloodstream through the intestinal **villi** (tiny, finger-shaped structures in the mucous membrane). Therefore, the blood passing through the capillary beds surrounding the intestines contains high concentrations of simple sugars, amino acids, and other absorbed molecules. It is imperative that the high concentrations of these compounds be brought rapidly to a lower level, or this blood chemistry could radically affect the well-being of the animal. The remaining intestinal contents pass to the **large intestine**, where water absorption occurs before the indigestible wastes pass through the terminal portion of the intestine, the **rectum**, and out the **anus**. These wastes are the **feces**, or excrement.

Immediately after passing from the stomach and the intestine, the nutritionally enriched blood goes to the liver, which removes most of the excess simple sugars, converting them to the insoluble polysaccharide glycogen. Converted simple sugars are stored in this more complex, less reactive form until they are needed. When the blood leaves the liver, it contains a concentration of glucose that is only slightly higher than the normal level found throughout the rest of the vascular system. This sugar is then taken up as it passes through the rest of the body.

The liver stores glycogen until the liver's storage capacity exceeds its limit. Then, any excess glucose gets converted into fat, which is stored in fatty tissues located throughout the body. Therefore, the liver helps to maintain a stable blood sugar level, regardless of how much carbohydrate is consumed. If a period elapses during which an individual is unable to obtain the proper carbohydrates, then the liver converts some glycogen into glucose and releases it into the bloodstream. However, the human liver stores only about 24 hours worth of glucose. When the total amount of glycogen stored in the liver is depleted, the body begins converting its stored fat into glucose.

In addition to the liver, the pancreas also helps to control the blood sugar levels via the hormones released. Insulin decreases the blood's glucose concentration, and glucagon increases the blood's glucose concentration.

ISOMERS

Some of the simple sugars have the same number of carbon, oxygen, and hydrogen atoms. But their atoms have different spatial arrangements, resulting in properties specific to each. Such compounds that are composed of the same kinds and numbers of atoms but are arranged differently, and therefore have different properties, are known as **isomers**.

Isomers can be classified into different categories. Some differ in the grouping of their atoms. For instance, a molecule may have an –OH group attached to the last carbon, making it an alcohol, or an oxygen may be bonded between the two carbons, making it an ether. Figure 14.2 shows the alcohol and ether isomers of C_2H_6O.

The examples in Figure 14.2 are called **structural isomers** because they differ in basic molecular structure. **Geometric isomers** also differ in the spatial arrangement of their atoms because the carbon-to-carbon double bonds cannot rotate, as they would be able to if they were single bonds. Two examples

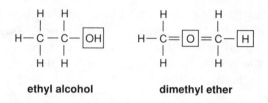

ethyl alcohol   **dimethyl ether**

Figure 14.2 *These two isomers of C_2H_6O are structural isomers.*

of geometric isomers are provided in Figure 14.3; both have the chemical formula $C_4H_4O_4$.

Another type of isomer can be explained in terms of a mirror image. While some mirror image isomers are functionally equivalent to one another, others represent molecular rearrangements that are distinct both structurally, in that they are arranged differently from one another, and chemically, in regard to how they react to each other. Two such **optical isomers**, as they are called, are presented in Figure 14.4; both have the chemical formula $C_3H_6O_3$.

Figure 14.5 shows examples of isomeric hexoses—the three monosaccharides with the chemical formula $C_6H_{12}O_6$. Although all three of these sugars are shown with straight carbon chains, they occur in aqueous environments with all six carbons attached in a ring. The ring structure of glucose is illustrated in Figure 14.6 on page 238. The other monosaccharides with chains of five or more carbon atoms usually become rearranged into the ring structure when they are dissolved in water, as is the case in living organisms. A graphic way to illustrate these structures, both for simple sugars and for more complex sugars, which are composed of strings of ring structures, is illustrated in Figure 14.7 on page 239.

When two simple sugars are bonded together, the product is a double sugar, or **disaccharide**. The bonding process is similar to that of amino acids described above. It involves the removal of a molecule of water and is called a dehydration reaction (or condensation reaction). Such reactions form the disaccharides sucrose, maltose, and lactose as well as **polysaccharides** such as glycogen, starch, and cellulose. The dehydration reaction involving two molecules of glucose is illustrated in Figure 14.8 on page 240.

After each monomer loses an atom or two, the remaining part is the **residue**. Condensation reactions are reversible; when a water molecule is

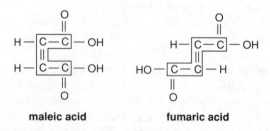

maleic acid   **fumaric acid**

Figure 14.3 *These two isomers of $C_4H_4O_4$ are geometric isomers; note the double bond between the two carbon atoms.*

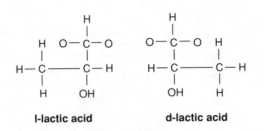

l-lactic acid   d-lactic acid

Figure 14.4 *These two isomers of $C_3H_6O_3$ are optical isomers.*

added between the monosaccharide residues, they will split by **hydrolysis**, which means water-breaking. The carbohydrate molecules that cannot be broken down into smaller units by hydrolysis are monosaccharides. By a series of condensation reactions, many monosaccharides can be joined into polysaccharides, some of which are of great biological importance. **Starches**, for example, are composed of long chains of glucose. Many plants store excess sugar molecules as starch. Another polysaccharide found in most animals is glycogen (see Figure 14.9 on page 240), which is similar to starch but has more branching sequences and is one of the main carbohydrates in animal tissues. It is made and stored in the liver and muscles, where it is available as an energy source.

Another biologically important complex polysaccharide is **cellulose**, which supports the bodies of higher plants. Like starches, cellulose is formed by **polymerization**, the bonding together of small molecules to form long chains. Both starch and cellulose are formed by large numbers of glucose rings that are bonded together, but the precise arrangement of the two polysaccharides differs slightly.

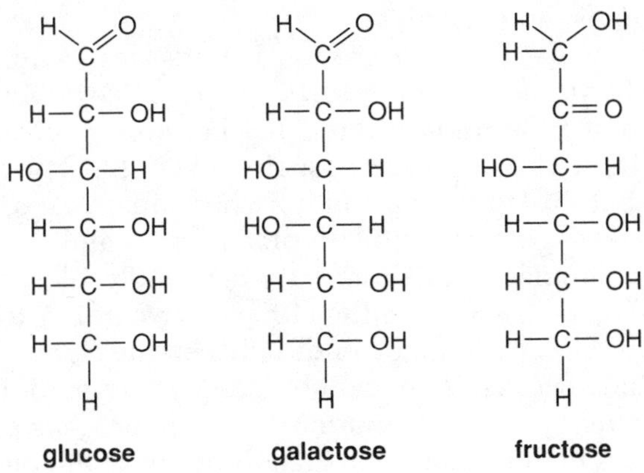

glucose   galactose   fructose

Figure 14.5 *Isomers of $C_6H_{12}O_6$; these three are isometric hexoses (6-carbons).*

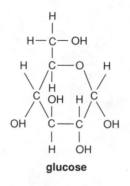

glucose

Figure 14.6 Ring structure of glucose ($C_6H_{12}O_6$).

LIPIDS

Other biologically important compounds are the **lipids**, which like carbohydrates are composed largely of carbon, oxygen, and hydrogen. They are defined by their solubility, being insoluble in water and soluble in nonpolar organic solvents such as ether, chloroform, and benzene. Because of their insolubility in water, lipids are vital in the composition of cell membranes. Lipids are also important because they store energy in a concentrated form without interfering with the body's water balance, unlike many energy-rich compounds such as the smaller carbohydrates that dissolve in water and therefore affect the cell's osmoregulation.

Proportionately lipids contain much more hydrogen and much less oxygen than do carbohydrates. Because of the many energy-rich carbon-hydrogen bonds, lipids have slightly more than twice the number of calories as equal amounts of carbohydrates. The major types of lipids are the **fatty acids, triglycerides**, **phospholipids**, and **steroids**.

Fats, the most widely known group of lipids, are composed of glycerol and fatty acid subunits (see Figure 14.10 on page 241). Glycerol is a three-carbon compound that is quite similar to the simple sugars. A fatty acid consists of a chain of carbon atoms, usually 14 to 22, with hydrogen atoms attached throughout except at the end, where there is a carboxyl group (–COOH). This makes the compound an organic acid and, in this case, a fatty acid. A glycerol combines with three fatty acid molecules, making fats and oils; the technical term usually used is either **triglycerol** or **triglyceride**.

Fatty acids without any double bonds between the carbon atoms are called **saturated fatty acids** because they cannot take up any more hydrogen atoms. That is, they are already saturated with as many hydrogens as possible. Such fatty acids are found mainly in animal fats. Those with double bonds between some of the carbon atoms can still take up more hydrogens if the double bonds are converted to single bonds. The adding of more hydrogens is called **hydrogenation**. Fatty acids containing some double bonds are called **unsaturated fatty acids**. They are found mainly in plants.

Carbohydrates		Sources
	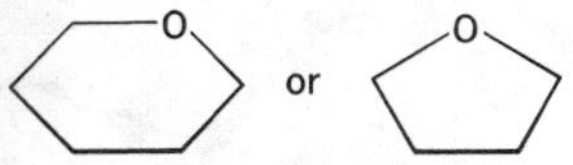 Simple sugars: or	Glucose Fructose Galactose Ribose	Animals and plants Fruits and honey Milk Nucleic acids
	Double sugars: 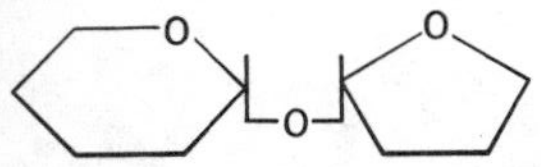	Sucrose (glucose + fructose) Lactose (glucose + galactose)	Sugar cane and sugar beets Milk
Complex carbohydrates	Multiple glucose units 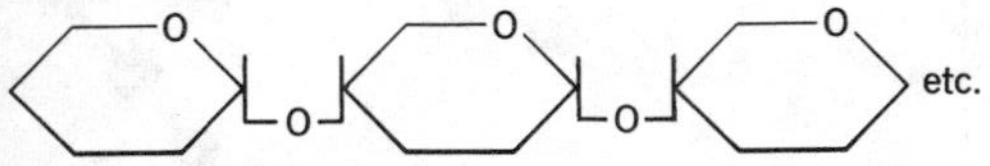 etc.	Glycogen Starch Cellulose	Liver and muscle Many plants All plants

Figure 14.7 Some common simple, double, and multiple molecules composed of sugar ring units, with examples and sources.

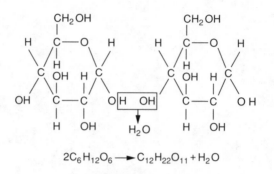

$$2C_6H_{12}O_6 \longrightarrow C_{12}H_{22}O_{11} + H_2O$$

Figure 14.8 *Sucrose, our common table sugar, is a disaccharide composed of a glucose and fructose unit. Cells use the energy in the chemical bonds linking the hydrogen and carbon atoms.*

Food labels listing the ingredients often term an oil or fat hydrogenated when hydrogen atoms have been added to the fatty acids to thicken the consistency of the product. Although this makes the product less runny, it also makes it less healthy. There is considerable evidence that saturated fatty acids are bad for you. There is also considerable evidence that unsaturated fatty

glucose unit

Figure 14.9 *Glycogen is composed of long branching chains of glucose molecules. Strings of glucose units are often stored in this form.*

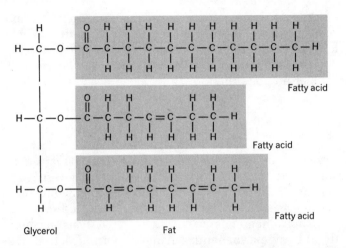

Figure 14.10 *Each fat molecule is composed of a glycerol unit to which are attached three fatty acid molecules (the fatty acid units are usually much longer than those shown here).*

acids help protect people from heart and arterial disease such as atherosclerosis, which results from the accumulation of fatty material inside the arteries. These deposits cut down the blood flow and can increase blood pressure. Should a piece of the fatty deposit break off, it might lodge in a blood vessel in the heart or brain, causing a heart attack or stroke.

The phospholipids are another important group of lipids. They are composed of glycerol, two fatty acids, and phosphoric acid, and in most cases a nitrogenous compound is also attached. The phospholipids are an important component of the cell's membrane.

The steroids are another large family of lipids comprising many hormones, vitamins, and other body constituents. They have four joined carbon rings. **Cholesterol**, notorious for being the substance that is laid down inside atherosclerotic arteries, is a steroid. Testosterone, estrogen, progesterone, and cortisone are four other important steroids. Synthetic steroids that mimic

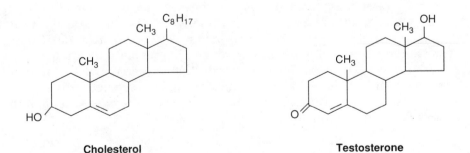

Figure 14.11 *Ring structures of two well-known steroids.*

the male sex hormone, testosterone, known as androgenic anabolic steroids, are used by many athletes to build muscle mass. Researchers feel these compounds are dangerous, and their use is prohibited by the International Olympic Committee. The illustrations of cholesterol and testosterone in Figure 14.11 show the basic steroid construction.

VITAMINS

Vitamins are organic molecules that are vital in small quantities to all living organisms. Many participate in biochemical reactions; for instance, some help to activate specific enzymes. Such vitamins are **coenzymes**. Other vitamins act as catalysts by assisting either in reactions that produce energy or in the manufacture of such chemicals as hormones. Catalysts accelerate chemical reactions without being permanently changed by the reaction. Other vitamins help activate reactions that synthesize tissues, break down waste products and toxins, or help remove toxic substances.

Vitamins are classified into two groups: fat soluble vitamins and water soluble vitamins. Water soluble vitamins, when consumed in the diet in quantities greater than required, are readily passed through one's system and excreted, mostly through the urine. These include the various B vitamins, such as thiamine and riboflavin, and vitamin C. The fat soluble vitamins are A, D, E, and K. Most vitamins can be extracted from plants and animals, being concentrated especially in their most metabolically active tissues. Table 14.1 provides a listing of the water soluble and fat soluble vitamins.

MINERALS

Although carbon, hydrogen, and oxygen are three of the main elements occurring in the organic molecules so vital to plants and animals, many other elements are necessary for chemical reactions and the construction of specific molecules. One example is magnesium which, though needed only in trace amounts, is an absolutely necessary component of chlorophyll. Likewise, iron is an important part of the hemoglobin molecule found in the red blood cells. It helps carry oxygen to the body cells. There are about 25 different minerals that are required for life to exist. Some of these are needed in relatively large amounts, while others are needed in only trace amounts. While the daily number of grams of iron typically equals about 18, only about 0.015 grams of zinc are required. Individual differences in people's requirements can vary from 20 to 50 times. Other minerals that are essential for maintaining human health include: calcium, chlorine, cobalt, copper, iodine, manganese, phosphorus, potassium, and sodium. Plants also require specific minerals above minimum and below maximum levels.

Table 14.1 *Water Soluble and Fat Soluble Vitamins.*

Name	Sources	Function	Deficiency Symptoms	Daily Need
Fat soluble Vitamin A	Dairy products, egg yolk, green and yellow vegetables	Healthy skin, resistance to infection, formation of visual pigments	Scaly skin, susceptibility to infection, night blindness	5000 IU*
Vitamin D	Fish, liver, milk, sunlight action on the skin	Calcium metabolism	Rickets, muscular weakness	400 IU
Vitamin E	Green, leafy vegetables	Involved in electron transport chain	Anemia, male sterility	15 IU
Vitamin K	Leafy vegetables	Blood clotting	Excessive bleeding	Unknown
Water soluble Vitamin B_1 (Thiamine)	Organ meats, whole grains, green vegetables	Involved with Krebs cycle, normal growth and metabolism, appetite, nervous stability	Beri-beri, irritability, loss of appetite	1.5 mg
Vitamin B_2 (Riboflavin)	Dairy products, meat, green vegetables	Normal growth and metabolism, healthy skin, part of electron carrier FAD	Dermatitis	1.8 mg
Niacin	Whole grains, meat	Healthy skin, cell respiration, part of electron carrier NAD	Pellagra, nervous disorders	20.0 mg
Vitamin B_{12}	Liver	Red blood cell maturation	Pernicious anemia	0.003 mg
Vitamin C (Ascorbic acid)	Citrus fruits, tomatoes	Ground substance in cells	Scurvy, anemia, hemorrhage	45 mg

*IU = International Units

FIBER

Much has been said about the importance of **fiber**, the indigestible part of the vegetable matter one consumes. Fiber consists in large part of cellulose and lignin, two polymers that human digestive systems are unable to digest. We get relatively large amounts of these materials in our diet when we eat vegetables, fruit, and cereals.

The value of fiber lies in the passage of our digesting food through the digestive tract. Peristalsis works best when there is much food in the intestine. However, if by the time everything we eat reaches our intestines, it is soft and without bulk or substance, then the residues will remain there longer than normal. The extra time promotes bacterial action which results in gas, diarrhea, or constipation. Research indicates that years of this type of digestion can significantly increase the chances of developing colon cancer.

Key Terms

alcohol group
aldehyde group
amino acid
amino group
amylase
anus
calorie
carbohydrate
carboxyl group
catalyst
cellulose
cholesterol
coenzyme
condensation reaction
covalent bond
deamination
dehydration reaction
digestion
duodenum
enzymes
esophagus
essential amino acid
ester group
fats
feces
fiber
fructose
functional groups
galactose
gall bladder
gastric juice
geometric isomer

glucose
hexose
hydrogenation
hydrolysis
inorganic molecules
isomer
keto group
kwashiorkor
large intestine
lipid
methyl group
minerals
monomer
monosaccharide
nucleic acid
organic molecules
optical isomer
pancreas
pentose
peptide bond
peristalsis
phosphate group
phospholipid
polymer
polymerization
polypeptides
polysaccharide
protein
pyloric sphincter
rectum
residue
saliva

saturated fatty acids
small intestine
starches
steroid
stomach
structural isomer
tetrose

triglyceride
triglycerol
triose
unsaturated fatty acids
villi
vitamin

Chapter 14 Self-Test

QUESTIONS TO THINK ABOUT

1. Why is it necessary for food to be digested?
2. Describe some simple sugars and explain their nutritional value.
3. Why are proteins important in a diet?
4. What is the role of lipids in one's diet?
5. What are vitamins and why are they important?
6. What is the difference between a vitamin and a mineral?
7. What is thought to be the value of fiber in one's diet?

MULTIPLE-CHOICE QUESTIONS

Proteins

1. Nutritional energy is measured in _____, one of which is equivalent to the amount of energy needed to raise one cc of water one degree centigrade.
   a. calories
   b. moles
   c. Kelvin
   d. amperes
   e. volts

2. Breaking down the materials containing nutritional value, which we call food, involves a series of processes that in the aggregate are known as _____ .
   a. digestion
   b. moles
   c. calories
   d. nutrition
   e. proteination

3. Proteins are made of _____ .
   a. monosaccharides
   b. disaccharides
   c. simple sugars
   d. cellulose
   e. amino acids

4. Amino acids contain one _____ and a carboxyl group.

    a. amino group
    b. monosaccharide
    c. simple sugar
    d. phospholipid group
    e. cellulose

5. Enzymes are a group of _____ .

    a. essential amino acids
    b. monomers
    c. proteins
    d. lipids
    e. a, b, and c

6. Most amino acids can be manufactured from smaller constituents, but others have to be taken in the diet already intact; these are called the _____ .

    a. enzymes
    b. catalyst
    c. proteins
    d. essential amino acids
    e. none of the above

7. A protein deficiency may be diagnosed as _____ .

    a. scurvy
    b. rickets
    c. polio
    d. kwashiorkor
    e. none of the above

Carbohydrates

8. Most diets consisting of just _____ and _____ can eventually prove fatal.

    a. fats and proteins
    b. fats and carbohydrates
    c. carbohydrates and proteins
    d. essential amino acids and fats
    e. essential amino acids and carbohydrates

9. Carbohydrates, with their same basic molecular arrangement, range in size from the smallest _____ to the larger _____ .

    a. simple sugars, polysaccharides
    b. monosaccharides, polysaccharides
    c. simple sugars, starch
    d. monosaccharides, cellulose
    e. all of the above

10. The length of the carbon chain in simple sugars varies, depending on the specific sugar involved. Some contain just three carbons; these are the _____ ; those with four carbons are _____ ; five carbons _____ ; and six carbon sugars are _____ .
    a. trioses, tetroses, pentoses, hexoses
    b. tetroses, pentoses, hexoses, trioses
    c. pentoses, hexoses, trioses, tetroses
    d. hexoses, trioses, tetroses, pentoses
    e. none of the above

11. Compounds having the same molecular weight and molecular formula as one another, or, in other words, having the same number of atoms of the same kind but arranged differently, are known as _____ .
    a. isomers
    b. tetroses
    c. hexoses
    d. triomers
    e. none of the above

12. When two simple sugars are bonded together, the product is a _____ .
    a. monosaccharide
    b. disaccharide
    c. trisaccharide
    d. quatrosaccharide
    e. none of the above

13. When two simple sugars are bonded together, the bonding process involves the removal of a molecule of water and is called a _____ .
    a. condensation reaction
    b. dehydration reaction
    c. hydrolysis reaction
    d. a and b
    e. a and c

14. After a dehydration reaction, the remaining part of each monomer is called a _____ .
    a. condensate
    b. dehydrate
    c. hydrolate
    d. monomate
    e. residue

15. Dehydration reactions are reversible. By adding a water molecule between the monosaccharide residues, they can be torn apart. Such a reaction is known as _____ .

   a. dehydration
   b. hydrolysis
   c. demonomerization
   d. a and b
   e. a and c

16. By a series of dehydration reactions, many monosaccharides can be joined into long chains, creating complex carbohydrates, which are known as

   _____ .

   a. disaccharides
   b. trisaccharides
   c. sucrose
   d. maltose
   e. polysaccharides

17. _____ is composed of long chains of glucose.

   a. sucrose
   b. maltose
   c. galactose
   d. fructose
   e. starch

18. A polysaccharide found in most animals is _____ , which has more branching sequences than starch, and is one of the main carbohydrates in animal tissues. It is made and stored in the liver and muscles, where it is available as an energy source.

   a. starch
   b. cellulose
   c. glycogen
   d. fibrinogen
   e. none of the above

19. A biologically important molecule that supports the bodies of higher plants and is a polysaccharide formed by polymerization, is _____ .

   a. starch
   b. cellulose
   c. glycogen
   d. fibrinogen
   e. none of the above

Lipids

20. A group of biologically important compounds that are defined by their solubility, being insoluble in water and soluble in nonpolar organic solvents such as ether, chloroform, and benzene, are known as _____ .

    a. carbohydrates
    b. proteins
    c. lipids
    d. a and b
    e. a and c

21. The major types of lipids are the _____ , _____ , _____ , and the _____ .

    a. amino acids, fatty acids, triglycerides, and steroids
    b. amino acids, triglycerides, steroids, and phospholipids
    c. amino acids, phospholipids, steroids, and fatty acids
    d. fatty acids, triglycerides, phospholipids, and steroids
    e. all of the above

22. Fats are one of the most widely known groups of lipids. They are composed of two different types of subunits: _____ and _____ .

    a. triglycerides and phospholipids
    b. fatty acids and steroids
    c. glycerol and steroids
    d. glycerol and phospholipids
    e. glycerol and fatty acids

23. Glycerol is a _____ carbon compound that is quite similar to the simple sugars.

    a. one
    b. two
    c. three
    d. four
    e. five

24. A fatty acid consists of a chain of carbon atoms usually from _____ , with hydrogen atoms attached throughout, except at the end where there's a carboxyl group.

    a. 0–4
    b. 4–10
    c. 10–14
    d. 14–22
    e. 22–50

25. Fatty acids without any double bonds between the carbon atoms are called
    _____ .

    a. unsaturated fatty acids
    b. polyunsaturated fatty acids
    c. saturated fatty acids
    d. polysaturated fatty acids
    e. none of the above

26. Those fatty acids with double bonds between some of the carbon atoms
    can still take up more hydrogens if the double bonds are converted to
    single bonds. The adding of more hydrogens is called _____ .

    a. bonding
    b. hydrogen bonding
    c. hydrogenation
    d. a and b
    e. a and c

27. Fatty acids containing some double bonds are called _____ .

    a. saturated
    b. unsaturated
    c. polyunsaturated
    d. polysaturated
    e. none of the above

28. There is considerable evidence that saturated fatty acids are bad for you,
    and there is considerable evidence that unsaturated fatty acids help protect
    people from heart and arterial diseases, such as _____ , which
    is the result of fatty material that is laid down inside the arteries.

    a. scurvy
    b. kwashiorkor
    c. polio
    d. rickets
    e. atherosclerosis

29. An important group of lipids composed of glycerol, fatty acids, and phos-
    phoric acid, and, in most cases, with a nitrogenous compound attached
    are known as _____ .

    a. triglycerides
    b. steroids
    c. phospholipids
    d. fatty acids
    e. none of the above

30. The _____ are another large family of lipids comprising many hormones, vitamins, and other body constituents. They have four carbon rings which are joined.

a. triglycerides
b. steroids
c. phospholipids
d. fatty acids
e. all of the above

Vitamins, Minerals, and Fiber

31. Some vitamins act as _____ by accelerating certain chemical reactions without being permanently changed by the reaction.

a. minerals
b. fats
c. steroids
d. hormones
e. catalysts

32. Vitamins are classified into two groups: the fat soluble vitamins and the water soluble vitamins. Excess fat soluble vitamins consumed in the diet become stored in one's fatty reserves rather than being passed through one's system. Examples of fat soluble vitamins are _____ .

a. A, B, C, and E
b. A, B, E, and D
c. B, C, D, and E
d. A, C, D, and K
e. A, D, E, and K

33. The following mineral is an important component of chlorophyll:

a. calcium
b. copper
c. cobalt
d. manganese
e. magnesium

34. The following mineral is an important component of hemoglobin:

a. iodine
b. chlorine
c. potassium
d. magnesium
e. iron

35. _____ refers to the indigestible part of the vegetable, fruit, and grains one consumes.

   a. minerals
   b. fats
   c. chlorophyll
   d. fiber
   e. hemoglobin

ANSWERS

1. a	10. a	19. b	28. e
2. a	11. a	20. c	29. c
3. e	12. b	21. d	30. b
4. a	13. d	22. e	31. e
5. c	14. e	23. c	32. e
6. d	15. b	24. d	33. e
7. d	16. e	25. c	34. e
8. b	17. e	26. c	35. d
9. e	18. c	27. b	

CHAPTER 15

Ecology

Ecology is a theoretical, quantitative study of organisms, populations, species, communities, and ecosystems. Both locally and globally, this highly integrative discipline incorporates information, processes, techniques, and data from related fields represented in the previous chapters, as well as inorganic chemistry, organic chemistry, physics, meteorology, geology, soil science, evolution, genetics, immunology, pathology, bioengineering, behavior, and natural history. While all disciplines are interdisciplinary to some extent, few are as interdisciplinary as ecology.

The study of the interactions of organisms, with each other and with the environment, is far more complex than we can fully comprehend. However, that doesn't stop us from trying. Ecology has rapidly grown from a pure science, primarily the study of species in their natural habitats, to a science involved with ecological principles and mechanisms. This has been, at least in part, a response to our immediate need for greater understanding about the interactions of organisms in their natural, as well as in what are often highly altered, habitats.

With human disturbances affecting ecosystems globally, we need methods to assess the potential effects of our actions. We also need to know what might be done to maintain healthy ecosystems. In addition to understanding the major, underlying ecological principles, we require a better understanding of how organisms live. Presented in this chapter is an overview of the primary ecological concepts that are now discussed in a number of introductory biology courses, as well as a number of media, including newspapers and television.

HABITAT AND NICHE

The physical environment affects organisms living there in many different ways. The **climate**, **weather**, and **microclimates** are influenced by **temperature**, **humidity**, **rainfall**, and other meteorological phenomena. Soils are extremely important in their effect on terrestrial habitats. In addition, aquatic habitats are also affected by salinity, oxygen and carbon dioxide in solution, pH, temperature, as well as available minerals and other nutrients. The amount of light penetrating through any environment is important, as are the seasonal fluctuations with regard to any of the above variables.

Just as aquatic organisms are adapted to specific sets of variables, certain plants grow only on special soil types, and some animals live only in areas where there are certain associated vegetation patterns. Air quality is also important.

An organism's **habitat** is the place where it lives. The variables mentioned above determine the characteristics of that habitat. An organism's **niche** is what it does. That is, defining an organism in terms of its role in the ecosystem describes its niche. Niches may be described using as many variables as one studies. For instance, a niche might be represented in terms of where an animal forages, when it forages, and what it forages on. This could be contrasted to the niches of related species living in the same general area to see if their niches vary according to these variables. The choice of variables one may study are practically limitless.

POPULATIONS AND COMMUNITIES

All the members of a certain species living in a specific area at a specific time are defined as a **population**. Each of the populations living in a given area constitutes a **community**. Any change in the population size is due to either birth, death, immigration, or emigration. Should births and immigrations exceed deaths and emigration, the population will grow. Likewise, the opposite is also true: If the total number of deaths and individuals emigrating from the population exceed the gain from births and immigrating individuals, the population will decrease in size.

The **density** of a population is calculated by counting the number of individuals per particular area (or volume). **Dispersion** is a parameter used for understanding the type of spacing of individuals within the population. The individuals are found to be distributed either **uniformly**, **randomly**, or in **clumps**. The latter two distribution patterns are more common than the first.

POPULATION GROWTH

Many factors affect the numbers of organisms in each population. Such numbers may vary from season to season or year to year. Some populations are growing in numbers while others are declining. Many remain quite stable, sometimes having built-in fluctuations. When there are no checks controlling **population growth**, it may grow on a logarithmic scale. When plotted on a graph, such a growth curve is often termed exponential or geometric (see Figure 15.1). A more controlled growth curve may be linear and is sometimes called arithmetic.

The rate of change in numbers of individuals in a population during a period of time is related to that species' innate capacity for increase or **biotic potential**. The actual rate of population increase at any given time may be calculated and is presented as a number that is called r. The biotic potential is higher for fruit flies than it is for people; therefore flies have a higher r.

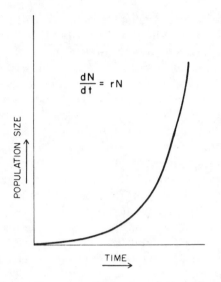

Figure 15.1 *Exponential growth curve. (N = population size; r = rate of increase; t = time; d = change.)*

A species' biotic potential is affected by any of the following:

1. The number of offspring produced each time the species reproduces.
2. When the reproductive life of that organism begins.
3. How long the reproductive life of that organism lasts.

In addition to representing survivorship with survivorship curves, **life tables** may also be constructed. These summarize, for each age class as defined by the researcher, specific types of information, such as the number of organisms living at the beginning of that age interval, the number of organisms that die during that age interval, and the average number of years of life remaining at the beginning of that age interval. Researchers often use this information to project what the effect on a population might be if certain variables are changed very slightly. Life table data have made it possible to project a range of effects without having to conduct nearly as much time-consuming fieldwork. Some computer programs have been especially useful in interpreting reams of data in new ways that were far too difficult just a few years ago.

SPECIES DIVERSITY

Ecological research has taught us about patterns in population size of individual species, as well as about systems involving the interactions of two or three species. But systems quickly become amazingly complex when more species are factored into the equation. There is still much to be learned about species interactions. **Ecological homeostasis**, defined as well-regulated species population sizes combined with **species diversity**, is affected by many components, only some of which are understood. Species diversity can be defined as the number and kinds of species present in any specific community. In addition to species diversity, the population size of each species is very significant in understanding the dynamics of that community. Some species are dominant. In fact, in most communities, there are only a few species with significantly greater abundances than all the other species. From this comes the often repeated phrase referring to numbers of species and their relative abundances: "few common, many rare."

What makes a species common in a particular community and what defines a "rare" species, are concepts currently being investigated by a number of researchers. One example of what has been learned is that there are **specialists** and **generalists**. That means there are some species that specialize at something. For instance, a particular lizard may be observed feeding on insects during a particular time of day, when the temperature is within a well-defined range, at a certain height in trees, on branches a certain width in diameter, eating only those insects that are within a narrow size range. There might be nine other closely related lizard species in that tree. However, each specializes in feeding at a different height, or on a different width branch, or at a different time, or on different food. Repeatedly we find situations where species, usually closely related, are "packed" together, but each species' niche overlap is minimal. This appears to be due to genetic control. Each of these species could be said to be a specialist. And then there are generalists, those species that might be moving all over the tree, as well as on the ground, throughout the day (or night, or both), eating a range of foods.

SPECIES TURNOVER, RECRUITMENT, AND INTRODUCED SPECIES

Another factor relates to **species turnover**. The species' composition within a community is not always stable. In fact, of those communities studied over time, it is common to find some species going extinct locally, a process usually referred to as **extirpation** and, concurrently, new species moving in (**recruitment**). The rate of species turnover in a particular community can help tell an investigator much about the local **interspecific** dynamics (dynamics between species, as opposed to **intraspecific** interactions, those within a species).

Many species are being introduced to areas where they are not native. Such exotic species, often referred to as **aliens**, may compete with native species for otherwise scarce resources. The net result of the increasing numbers of introduced species has been a growing list of native species that are decreasing in numbers and range. Some may eventually become extinct.

COMPETITION

When two members of one species or of different species exploit a common resource, the interaction is often termed **competition**, though the behavior is rarely observed. Designing experiments that enable a researcher to show whether or not competition exists and, if so, to what extent, usually requires great care and a rather simple system.

Competition may occur between different species (interspecific competition) or within a single species (intraspecific competition). When each individual directly competes for a resource without regard to other individuals that may also be using that resource, it is known as **scramble competition**. In cases where individuals compete indirectly for what may not be a very tangible resource, such as for social dominance or for territory, it is often called **contest competition**. Here, successful individuals usually get enough of the limited resource while the losers often go without.

The **competitive exclusion principle** states that when two or more species compete for the same resource necessary for their livelihood, and they are found or placed together in a situation where the resource is limited, they cannot coexist indefinitely. One will win out in the long run. Put another way, two species cannot occupy the same niche.

Sometimes the niche that a species might occupy differs according to the situation. That is, when only one species is there, the **fundamental niche** (what the species does when there is no competition) may be larger than the **realized niche**, the niche that is occupied when the environmental conditions are more restricted because one or more species are infringing on that otherwise larger, more expansive niche.

DENSITY

What causes the increased mortality of a species is not always clear-cut. As the population density of one species increases, the mortality within that species may increase. In such a case the mortality is said to be **density dependent**. However, if the mortality is due to causes other than the species' density, then it is said to be **density independent**.

CARRYING CAPACITY

The **carrying capacity**, often referred to by the letter K, is the number of individuals of a particular species that a specific environment can support indefinitely. Change the environment, and K changes. Add a parasite, remove a competitor, add a predator—each of these factors may affect the carrying capacity of the species in question. The curve often used to show the exponential increase in numbers of a species until it reaches its carrying capacity in that environment is called a **logistic curve** (see Figure 15.2).

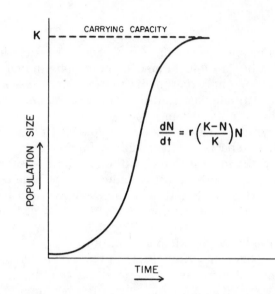

$$\frac{dN}{dt} = r\left(\frac{K-N}{K}\right)N$$

Figure 15.2 *Population growth as described by the logistic equation. (K = carrying capacity; other variables as defined in Figure 15.1.)*

SUCCESSION

The population size of each species and the species diversity in many communities are stable and persistent, often remaining predictably uniform for centuries or more. Other populations are far more fluid, or dynamic, entities in the constant process of change from what may appear to be one distinct community to another. Such a sequence of communities replacing one another is known as **succession**, which is usually far more gradual than is commonly recognized. Successional events that occur on land are known as terrestrial succession; successional events that occur underwater are known as aquatic succession.

Succession may be triggered by a cataclysmic event such as a flood, fire, or even a plow or bulldozer. But sometimes, the changes are triggered by a far subtler influence, such as a disease, a slight increase in average summer temperatures, or the invasion of a new insect.

The exact species composition is the result of such factors as soil, climate, which seeds arrive first, and which seeds germinate first. When a cataclysm is initiated by a plow, for instance, seeds germinate on the cleared soil. Some seeds are blown in, others deposited in animal feces, so which animals live nearby can be as important as any other single factor. The initial colonizers also impose certain characteristics: They act as a filter, enabling only certain new species to prosper.

The field may be composed of grasses and herbs, annuals, biennials, and perennials. The precise mix changes with time, and eventually woody species take hold and shade out many of the field species. In addition to the woody shrubs, pines and cedars are good colonizers that grow quickly. However, if fires don't burn them back, perhaps every 15 years or so, the evergreens may soon be overtaken by deciduous hardwoods, species such as oaks and maples. The reason pines may not be overtaken by hardwoods if fires persist with regularity is that the pines are often fire resistant. Although they suffer considerable damage, they also grow back quickly, and in a number of years the area will once again be a pine forest. But without the fires, the hardwoods overtake the softwoods (pines) because they grow higher, shading out and eventually causing the pines to die off.

In this continuum of different communities that gradually change into one another, the concept of the final stable community is often evoked. That is, in some places, an oak-hickory forest might be the last natural successional community. Elsewhere it might be a beech-maple forest, depending on the weather, soils, and a number of other factors. Other final, stable communities, also known as **climax communities**, are deserts and grasslands. Climax communities occur only after the gradual series of continual changes. Preceding communities are lost to better adapted competitors that thrive under these different conditions, made different by the new plants and animals living there.

Once in place, a climax community lasts until the next flood, fire, or bulldozer comes along, when the successional processes are reinitiated.

REPRODUCTIVE STRATEGY

Natural selection favors organisms that tend to pass on more offspring, which in turn pass on more offspring, and so contribute to succeeding generations. Included among the characteristics that natural selection appears to affect are those that shape an organism's **reproductive strategy**. Such strategies are not conscious. Rather, they evolve over time. There are specific categories of reproductive strategies, with many variations. For instance, some organisms produce great numbers of minute offspring, while other species invest more in each offspring; they have only a few, but each is large, with considerable survival value. One of the primary differences among most reproductive strategies is the survivorship of the young. If each parent can have just a few young but still pass on enough offspring to maintain the population, that might be sufficient. However, if there is extremely high mortality, and only a very small percentage of the young survive to maturity, it is sometimes best to bear very many young, investing little in each. Each species has its characteristic reproductive strategy.

Recording and plotting the fate of the young and their chances of survival at key age categories enables the researcher to construct **survivorship curves**. **Birth rates**, **death rates**, and **longevity** (how long the organisms may live) are significant factors affecting all populations. Recording the numbers of births and deaths over a period of time and determining the average longevity of the organisms during each age class, tell the scientist much about that population. Should the number of births rise, the number of deaths fall, or the survival time of the organisms increase, the population will be affected positively; that

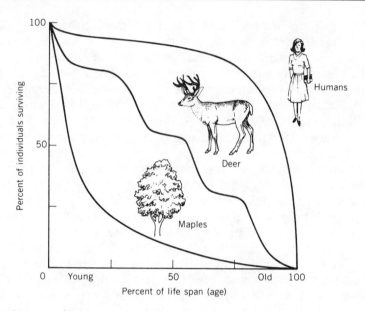

Figure 15.3 *Survivorship curves for humans, deer, and maples.*

is, the population should grow. Conversely, should the births decrease, the death rate increase, or the average longevity decrease, the population would be affected adversely.

Average longevity varies from species to species. When the survivorship curves are plotted, much can be determined about the reproductive strategy of that species. The three basic curve shapes are presented in Figure 15.3, although there are many variations.

A Type I survivorship curve (e.g., humans in Figure 15.3) depicts a species that has offspring with high survival rates; most live to a certain age and then die. A Type II survivorship curve (e.g., deer in Figure 15.3) depicts organisms with a steady death rate stretching from the time they are born or hatch until the time they die. Their survivorship curve usually varies along a straight line. And a Type III survivorship curve (e.g., maples) is the inverse of a Type I curve; it depicts a low survivorship shortly after being born, but with high longevity for the individuals that survive.

ECOSYSTEMS

Ecosystems are the sum total of the organisms and their environment in a given area; all require energy. The sun is the ultimate energy source in most ecosystems. The light energy is used by plants to convert, during photosynthesis, carbon dioxide and water into carbohydrates. Plants also use additional inorganic nutrients and ions to manufacture nucleic acids, proteins, photosynthetic pigments, and other necessary components.

Organisms such as plants, which manufacture their own organic food molecules from inorganic substances, are the **producers** and are often called **autotrophs**. Organisms that acquire food from dead plants and animals are called **decomposers**. In some ecosystems, the metabolic pathways of certain bacteria use chemicals rather than sunlight to synthesize organic materials. Such alternative methods of manufacturing organic food molecules are termed **chemosynthesis**.

Ecosystems require an energy source, producers, decomposers, and **abiotic** (nonliving) nutrients. These are just the basics; most ecosystems have more components. When there are organisms present that eat the plants, such animals are called **herbivores**, or **primary consumers**. When these die, like all dead material, they are broken down by decomposers. However, when living herbivores are eaten, their **predators** are termed **carnivores** (animals that eat other animals or animal parts); they are the **secondary consumers**. Depending on the complexity of the particular ecosystem, sometimes there are **tertiary**, or even **quaternary consumers**.

The autotrophs, or plants, represent the first level in the series of organisms that together are sometimes thought of as creating a **food chain**. Because of all the connections and complexities within most food chains, they are also called **food webs**. The plants are the first and most basic form of energy storage in an ecosystem. They are called the first **trophic level**. The second contains those animals that eat the plants, and the third, when present, consists of animals that eat herbivores. Organisms that eat both plants and animals belong to more than one trophic level and are called **omnivores**.

When the energy is passed from one trophic level to another through the consumers, a certain percentage of energy is lost through the processes involved in hunting, eating, digesting, reproducing, growing, and maintaining. There is less energy available in each succeeding trophic level, which explains why few ecosystems contain more than five trophic levels. Beyond that, rarely is there enough energy to support much else.

Sometimes the fewer numbers of organisms in each succeeding trophic level are illustrated as a pyramid, with the earlier trophic levels at the base, building up to the narrow peak. Not only is a pyramid of numbers helpful, but a pyramid of biomass is often used to depict the total dry weight of living material in each succeeding trophic level. Depending on what is occurring in any particular ecosystem, occasionally the pyramid is inverted, with the primary producers representing either fewer numbers or less **biomass** and the larger base on top. Such situations are usually short-lived.

The energy accumulated by the plants in a given ecosystem is known as the **primary production**. The rate at which energy is stored as organic matter due to photosynthesis is called **primary productivity**. This stored energy is usually expressed as $kcal/m^2/year$, or as biomass gained per unit area over a given unit of time. Since some of the energy is immediately metabolized to maintain the plant's respiratory activities, it is sometimes helpful to calculate the **gross primary productivity**, which is the total rate of photosynthesis. Therefore, the gross productivity minus the respiration rate (energy used for respiration) leaves the **net productivity** (see Figure 15.4). It should be remem-

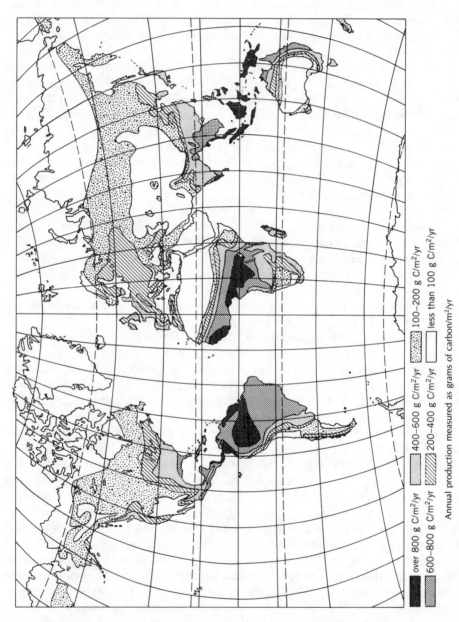

Annual production measured as grams of carbon/m²/yr

over 800 g C/m²/yr   400–600 g C/m²/yr   100–200 g C/m²/yr

600–800 g C/m²/yr   200–400 g C/m²/yr   less than 100 g C/m²/yr

Figure 15.4 *Annual net primary productivity of natural vegetation.*

bered that productivities are rates, yields are weights (or expressed as volume), and the **standing crop** is the total amount of biomass per unit area at any given time.

While primary productivity is the rate of energy storage for autotrophs, it is the **heterotrophs** that are involved in **secondary productivity**. This is the rate of formation of organic matter in the heterotrophs. Herbivores consume only about 1 to 3 percent of the total net primary production, although in some environments the percentages may be several times higher.

BIOMES

The major ecosystems, often called **biomes**, are regarded as distinct entities with distinct forms of life. These regional ecosystems generally grade into one another, forming a **gradient** or **transitional zone**. When the transition from one ecosystem to another is abrupt, leaving a significantly different environment along the barrier between the two, such as the beach or coastal zone that lies between the ocean and land, such a zone is sometimes called an **ecotone**. The major biomes of the world are depicted in Figure 15.5 in terms of their mean annual temperature and precipitation. Each of these ecosystems is described below.

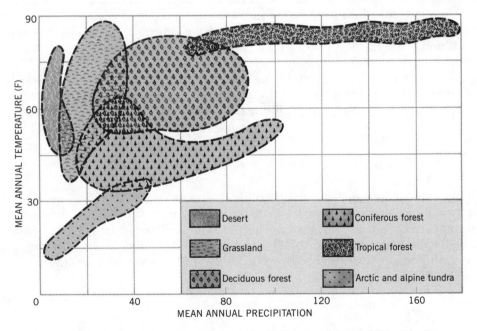

Figure 15.5 *Biomes are major ecosystems with distinctive forms of life. The biomes graphically represented here are presented in terms of temperature and precipitation.*

Tropical Rain Forest

Characterized by an annual rainfall of more than 400 cm (over 150 inches) and warm temperatures during most of the year rather than just during a specific warm or rainy season, **tropical rain forests** are wet and green all year. The soils tend to be nutrient-poor because the warmth and moisture have oxidized most of the available nutrients from the soil, which are then leached out by all the rain. Because the conditions are warm and moist, decomposers consume the organic material practically as soon as it becomes available. There is usually only a very thin layer of rich organic material on top of the forest floor and very little topsoil.

In terms of the overall number of species, or species diversity, the tropical rain forest is unsurpassed. These forests are most extensive in the northern part of the Southern Hemisphere, where they are being logged rapidly. In many cases it is local governments and international corporations that favor logging these forests for the short-term profits despite the long-term losses, should most of these forests disappear. In addition to the extinction of hundreds of thousands of species, it has been predicted that the loss of these forests will contribute to global warming.

Tropical Seasonal Forest

Unlike the tropical rain forests, which are largely evergreen since most of the trees retain leaves year round, **tropical seasonal forests** are composed of more **deciduous** tree species, due to more prolonged dry seasons. The dry seasons are similar to winter in colder climates, except for the lack of cold. This is because during the dry periods, and during the cold periods, many species become relatively dormant, and the majority of tree species lose their leaves. When the rains return, the plants leaf out and the organisms become more active.

Tropical Savanna

Typical and extensive in Africa, **tropical savannas** occur where the rainfall is capable of supporting vast expanses of grass with some trees and shrubs, but never any forests. Because of limited rain and the prolonged dry season, woody plants do very poorly. The wet season supports grasses that dry out the rest of the year. Some hardwoods persist in the lower areas with whatever runoff there is.

Tropical Thornwood

Where the rainy season is sufficiently wet to support some semidesert scrub, but conditions are too dry to support a forest, sometimes there are zones of **tropical thornwood**.

Desert

Because most **deserts** receive less than 20 cm of rain a year, their vegetation is sparse, and some deserts have little to no vegetation. Other deserts have many specialized plants that store water in their thick, fleshy tissues. Such plants are **succulents**.

Temperate Forest

Unlike the tropical areas, temperate climates have more moderate temperatures. Here, forests grow that are classified either as **temperate forests** or as **temperate deciduous forests**, having warm summers and cold winters. Like tropical seasonal forests, as well as other seasonally cold or arid ecosystems, most trees in this biome lose their leaves as the weather grows colder and grow new ones as the weather warms up. Soils are rich in organic matter. Despite richer soils, temperate forest trees rarely reach the heights typical of those in the tropical rain forests.

Temperate Evergreen Forest

Temperate evergreen forests grow on poor soils in areas where drought and fire frequently occur. The species that dominate these forests are usually needle-leaved conifers, such as pine, fir, and spruce, as well as some broad-leaved evergreens.

Temperate Rain Forest

Temperate rain forests prevail in cool areas near the ocean where there is considerable rainfall during the winter and much fog during the summer, such as the rain forests of the Pacific Northwest.

Temperate Woodland

In the American southwest, between the grassland and semidesert biomes at the lower elevations and the pine forests at higher elevations, are temperate woodlands that are too dry to support a forest but too moist for mere grassland. Instead, they support a range of communities from rather dense woodlands of small trees to more open woodlands with taller but more isolated trees.

Urban Ecosystems

Urban and suburban ecosystems are the newest, fastest-growing distinct ecological entities. And yet, these are the most overlooked habitats despite their far-reaching effects. New to much of the world, few urban environments have been in existence for more than 5,000 years. During the interim, vast areas have been transformed. Urban ecosystems are responsible for subsuming a range of other habitats into what now constitutes part of the meshwork of urban communities, including agricultural areas, which affect nearby rivers, estuaries, and even oceans. Because of gases released by activities of humans, the dominant organisms inhabiting these ecosystems, the atmosphere has been altered so other environments are suffering such consequences as global warming, rising sea level, and acid rain.

As the world human population grows, we continue to destroy healthy, "natural" habitats, replacing them with urban habitats, which in some respects are also "natural." In these habitats we find many species do quite well, not in spite of us, but because of us. We have created an environment where far more than rats, roaches, and pigeons do well. Often, we find such species, which we now associate with urban ecosystems, doing better than they did before we created cities, towns, and suburbs. The world is currently undergoing one of the major ecological changes of all geologic time, and the primary ecosystems to benefit, often at the expense of other ecosystems, are urban ecosystems. In the process, world species diversity is expected to decline, perhaps precipitously.

Temperate Shrubland

The **temperate shrubland** communities are often called **chaparral**. These relatively dry areas are found near water, such as along parts of coastal California or the Mediterranean. These areas get little to no rain during the summer and are subject to frequent burning. After the fires, the shrubs grow back, regenerating from the roots and tissues left near the ground.

Temperate Grassland

Depending on the country, the **temperate grasslands** are known either as prairie, steppe, pampas, or veldt. Located on vast lands in the interiors of continents where there is enough moisture to support shrubs and grassland, but not enough rain for a forest, temperate grasslands usually have rich deep soils, which are often used to support productive agriculture. Much of America's Midwest and the Soviet Union's Ukraine were temperate grasslands that are now farmed intensively.

Temperate Desert

True deserts are not very common in temperate regions. Such deserts are usually too dry for much life to exist. Rather, most **temperate deserts** are classified as semideserts, or as cool temperature semideserts. Too dry to support grasslands, they are often dominated by shrubs, such as sagebrush, with many perennial grasses throughout.

Boreal Forest

The **boreal forest**, often called the **taiga**, stretches through much of Canada and Siberia. This biome is dominated by spruce trees, pines, and firs that do well under what would be harsh conditions for most other species; it's cold and windy and the soils tend to be rocky and poorly developed. To survive here, a species must be able to tolerate the harsh winters.

Tundra

Tundra conditions exist farther north than the boreal forest, as well as at high elevations, where it is too cold for most trees to grow. Vegetation above the treeline primarily consists of sedges, grasses, mosses, lichens, herbaceous flowering plants, and dwarf woody shrubs.

Both environments share several characteristics, including shallow soils and a brief growing season. In some tundra regions the soils remain frozen all year round, a phenomenon called **permafrost**. Where this occurs, sometimes the surface thaws, but just below the surface temperatures never rise above freezing. Tundra plants grow at a comparatively slow rate. In addition, because of the cold weather, dead material decomposes at a rate that is many times slower than that in most warmer environments. **Alpine tundra**, unlike the true tundra of the north, is similar, except that these communities are found on mountain tops between the **timberline** and the highest areas, where little survives.

Streams and Rivers

These systems are typified by a current and fresh water. But beyond that, streams and rivers vary considerably. They range all the way from a roaring torrent pounding down a snowcapped mountainside, to the trickle of a dried-up summer brook, or an immense, meandering river such as the Amazon, working its way through the vast region of tropical rain forest in South America.

Ponds and Lakes

As with streams and rivers, most ponds and lakes have fresh water in common (though some are saline). They occur, however, in a wide range of temperatures, nutrients, typical species, and a considerable variety of associated habitats. There are different categories of ponds and lakes, but most **limnologists** (specialists who study such freshwater environments as lakes and ponds) find it helpful to group these freshwater bodies together. The experts, however, have a vocabulary that enables them to describe particular freshwater habitats with precision.

Freshwater Wetlands

Freshwater wetlands also vary considerably. The domain of most wetland biologists ranges from open water to wet meadows (which may be wet only during the rainy season). Specific vegetational patterns are associated with each type of wetland.

MARINE COMMUNITIES

Although the major biomes presented in most texts are usually restricted to terrestrial communities, there are many **marine communities**, those aquatic communities with more salt than fresh water. Like many of the communities presented above, marine communities also tend to merge into one another, but it is possible to identify distinct zones. Some of these are presented below.

Estuaries

The region where fresh water from streams and rivers drains into salt water is called an **estuary**. These coastal areas vary in salinity depending on the geography and the tides. The constant mixing and influx of nutrients, combined with the many protected habitats that typify these estuarine environments, make them one of the most productive ecosystems worldwide.

Estuaries often have **tidal marshes**. These are coastal areas with typical vegetational patterns. The dominant plant of salt marshes is *Spartina*, which is a genus, or group, of closely related grasses. *Spartina*, as well as many aquatic species inhabiting estuaries, can tolerate a range of saline conditions from several times the salinity of the ocean, which is 34 parts per thousand, to fresh, which has no salt at all. Pools on top of the salt marsh, due to evaporation, can become several times more saline than the open ocean. It should come as no surprise that of those organisms that inhabit the estuarine environment, most are well adapted.

These rich, protected environments are the breeding grounds for a wide range of invertebrates and fish, as well as species such as colonial water birds that feed on both of these. Many migrating birds and fish arrive in schools, sometimes just to feed. Others remain long enough to mate and lay eggs, and birds often spend an entire summer.

Littoral Zone

The intervening area between where the water reaches during high and low tide is known as the **littoral zone**. Many of the organisms inhabiting this zone can tolerate major swings in temperature and salinity, as well as the periods of desiccation when left out of water during low tide. Depending on the local conditions, littoral zones may be either rocky, sandy, or muddy. Each supports very different communities.

Coral Reefs

Coral reefs, along with estuaries and tropical rain forests, are among the most productive ecosystems in the world. Coral reefs occur in warm, shallow marine waters around the world. Usually confined to the tropics, coral reefs share several characteristics. They are all built by coral polyps (phylum Cnidaria) which secrete the hard, limy, calcareous material that becomes the infrastructure for all the other associated organisms essential to a healthy reef. Reefs may be exposed at low tide, but the majority of the coral lies under water. Coral reefs grow in clear, oxygenated waters where waves break. Coral species grow in water less than 100 meters deep because they can only exist where sufficient light reaches the associated, photosynthesizing organisms, such as the symbiotic species of algae.

Open Ocean

Of the other marine communities, the largest is found in the open ocean. Past the **sublittoral zone** that occupies the **continental shelves** (the edges of the continents where the underwater terrain gradually drops off), the continental masses drop off rapidly and, in some places, the ocean becomes very deep. In the clearest water, light can penetrate the first 100 meters. Beyond that, little to no photosynthesis is possible. Organisms, however, continue to live where there is no light, even as far as the deepest marine canyons that reach several kilometers in depth. The small creatures that make up the **plankton** are those that drift about in the water. Such species include algae and other organisms that are so small they are supported by the water's buoyancy. The larger animals, such as most mature fish, make up the **nekton**. These species usually have more directed, controlled movements.

MINERAL NUTRIENT CYCLES

Within each of the ecosystems described above are specific biologically active minerals that cycle through the plants and animals in that environment. These cycles are known as **biogeochemical cycles**. Unlike solar energy, the elements that function as nutrients in ecosystems can be recycled and are used over and over again in living organisms. The major biological nutrients are carbon, hydrogen, oxygen, nitrogen, and phosphorus. Other elements also prove important in lesser amounts; these include boron, calcium, chlorine, cobalt, iodine, iron, magnesium, manganese, potassium, and sodium.

Many biologically active elements emanate from rocks as they erode, or from the soils they help form. They are taken up through roots or are absorbed from water or through the air. The minerals may be incorporated into an organism's cellular structure, where they are used and then released back into the system, or they may be incorporated into the organism's tissues, to be released upon death and decomposition. Either way, the minerals eventually become available again where they can be absorbed and used by other organisms.

Carbon Cycle

The **carbon cycle** passes carbon from the atmosphere to organisms and back through respiration and photosynthetic processes. Most carbon is released into the atmosphere in the form of carbon dioxide (CO_2). In water carbon may exist either as dissolved carbon dioxide or as bicarbonate (HCO_3^-). Gradually, over hundreds of millions of years, considerable amounts of carbon have accumulated in some systems in the form of peat, coal, oil, or natural gas. Such carbon reservoirs can remain locked up for millions of years before rejoining the carbon cycling system again.

Since the beginning of the industrial revolution, the burning of fossil fuels has released much of this previously locked-up carbon into the atmosphere in molecules of carbon dioxide. All this carbon dioxide has led to the **greenhouse effect**, which is a warming up of the atmosphere because the added CO_2 in the atmosphere prevents some of the earth's heat from radiating back into space. Although the CO_2 concentration in the atmosphere has increased significantly during the past 20 years, it has increased only half as much as would have been expected from the burning of fossil fuels because much is absorbed by the oceans. Still, the total amount appears to have been sufficient to cause a slight increase in global temperatures.

All it takes is an average rise of a few degrees globally to alter weather patterns radically and melt portions of the polar icecaps. This would lead to a substantial rise in sea level, submerging today's beaches, salt marshes, deltas, and urban centers situated at sea level. In addition, the damage that could be expected during hurricanes, which regularly hit coastal cities and towns, would be devastating to the millions of people affected.

Nitrogen Cycle

Amino acids and proteins require nitrogen. However, even though there is far more nitrogen than carbon dioxide in the atmosphere (carbon dioxide = 0.03%; molecular nitrogen = 78%), most of this nitrogen (N_2) is in a form that can be used by only a very small group of specialized organisms.

Certain **bacteria** and **blue-green bacteria (cyanobacteria)** can convert, or fix, gaseous (molecular) nitrogen into an aqueous form, ammonia (NH_3) by reducing it. This means adding electrons, a process usually accomplished by combining the nitrogen with hydrogen, which carries an electron with it. It is primarily from these **nitrogen-fixing microorganisms** that most gaseous nitrogen is converted to a form capable of being used by other organisms. However, some nitrogen exists as ammonium ions (NH_4^+) or nitrate ions (NO_3^-) which, having been eroded from rocks and leached from soils, are dissolved in the waters around the world.

Certain bacteria and **fungi** use the nitrogenous materials found as proteins and amino acids in dead organisms for their metabolic purposes. They then release ammonia and ammonium ions back into the soil. Other bacteria in the soil, known as **nitrifying bacteria**, oxidize (remove electrons, usually by removing hydrogens, which carry the electrons with them) ammonia to nitrate ions (NO_2^-). Other bacterial species have the capacity to oxidize the nitrite ions to nitrate ions.

Lightning also produces nitrate. All these sources of fixed nitrogen are necessary to the rest of the organisms found around the globe because, without fixed nitrogen, there would be far less plant life and, without plants, little else could survive.

Because fixed nitrogen is so important to plants, it is a key ingredient in many fertilizers. Since it is expensive, farmers often choose to rotate their crops rather than fertilize, occasionally growing alfalfa or any of several other members of the pea family (Leguminosae), all of which have symbiotic nitrogen-fixing bacteria living in nodules attached to the roots. By growing such a crop, not only is the alfalfa harvested for horse and cattle feed, but fixed nitrogen is manufactured by the symbiotic bacteria and released into the soil, free of charge. The symbiotic nitrogen-fixing bacteria live in close proximity to legumes (plants in the pea family) so that they may receive nutrients from the plants. In exchange, the host plants benefit from the fixed nitrogen fertilizer. **Denitrifying bacteria** live under anaerobic conditions (no oxygen), such as exist deep in the soil or in mud. They complete the nitrogen cycle by returning nitrogen back to the air.

Phosphorus Cycle

Phosphorus is indispensable to life since it is an important component of ADP, ATP, RNA, and DNA. It is also found in cell membranes, as well as in shell, bone, and teeth. The previous two cycles, carbon and nitrogen, are termed **atmospheric cycles** in contrast to the **phosphorus cycle**, which is a **sedimentary cycle**. Carbon and nitrogen occur as gases during part of their cycles; phosphorus almost never does. Rather, through erosion it is generated from rock in the form of phosphate (PO_4^{3-}). Plants then absorb this phosphate. When animals eat the plants, or other animals, the phosphorus is in a form that is usable to them. Any excess phosphorus excreted by the animals is in the form of phosphate, so it may be reabsorbed by the plants.

Although much of the phosphorus stays in terrestrial ecosystems, a certain amount that originated from rocks and was absorbed by plants, which were eaten by animals, eventually gets washed downstream and out into the ocean. Eventually it settles to the ocean floor, where it remains for what may be millions of years before being exposed and eroded out, becoming part of the cycle again.

Much phosphorus used to be mined in the form of bat and bird droppings, the guano which accumulates in bat caves or on some oceanic islands where thousands of sea birds have nested for generations. Other sources exist, such as in local deposits in Florida, but these are being rapidly depleted. Because the supplies are limited, and those that persist will last only as long as it takes for worldwide agribusiness to buy and use them, phosphorus is an increasingly expensive fertilizer.

BEHAVIORAL ECOLOGY

Another integral part of ecology is the study of **animal behavior**, sometimes called **ethology** (or **biopsychology**). This field is concerned primarily with the investigation of what animals do in natural environments, particularly with regard to adaptation, natural selection, and evolution. More and more, plants, too, are being viewed in terms of their behavior. An organism's behavior is often related to its physiological features such as size, weight, and color, as well as its structure and the function of its parts.

The selective pressures that help shape each species' characteristics include the environment, food, predation, and reproduction. There are adaptations found among each species that have to do with all of the above. Understanding these adaptations helps one to understand ecosystem complexity.

Ecology and behavior, as well as evolution, are all closely related. The study of one invariably includes the others. For instance, in the discussion of the competitive exclusion principle in this chapter, the concepts of niche, habitat, predation, dominance, species diversity, and community dynamics were all invoked. **Symbiosis**, **commensalism**, **parasitism**, and **mutualism**, all of which have to do with behavior, are discussed below, with brief definitions of each of the major terms.

Most animals have an area in which they spend the majority of their time. This includes where they may interact with other members of the same species, look for food, rest, and sleep. Such an area is called a **home range**. This differs from an animal's **territory**, which is a subset of its home range. A territory is just the area that is defended, usually from the other members of the same species.

Not all organisms, but many, are social, and their social behaviors vary from species to species, and often from population to population. Some animals have **social hierarchies**, relationships in which the status, or position, of each animal with regard to the other members of the group is well established, either by some aggressive behavior, or just by relative size, age, sex, or a combination of characteristics. The dominance-subordinance relationships may change over time as animals grow, age, develop, mature, or die, leaving new openings in the hierarchy.

Some social aggregations amount to more than just simple "pecking-orders" of dominance-subordinance relationships. While certain species have very little, if any, division of labor within their social structure, other species have populations that function as a **society**. A society is a long-lasting arrangement of animals in which individuals are constantly changing, but the continuity of the social structure remains intact.

Symbiosis is another arrangement whereby species interact in a much more intimate way. It sometimes means living together, but most authors use it in reference to an interaction from which both species derive a mutual benefit. Normally, symbiosis is broken into three categories:

1. **Mutualism**—an arrangement between species in which each benefits from one another and neither is harmed by the relationship.

2. **Commensalism**—one species benefits while the other is neither harmed, nor benefited by the relationship.
3. **Parasitism**—one species benefits and the other is harmed.

Generally, the only difference between parasitism and a predator-prey relationship has to do with size. Often parasites are thought of as small animals that either live on or in their host. Those that live on a host are **ectoparasites** and those that live inside their host are **endoparasites**.

Predation, the feeding of free-living organisms on other free-living organisms, is one of the main forces affecting species adaptation and change over time (both behavioral change during a lifetime and genetic change over time). Organisms that reduce their chance of being eaten may accomplish this through self-defense, escape, or hiding. Generally the net positive effect is that they increase their chances of survival and, therefore, of passing on offspring to the next generation.

The different types of defense are **physical defense**, such as that employed by thorns or spines, and **chemical defense** such as chemicals that render the individual bad tasting or poisonous. With plants, if a chemical defense is effective, perhaps only one leaf is eaten. Many animals also have chemical defenses, which may be advertised with bright warning coloration; **aposematic coloration** warns potential predators to avoid these bad-tasting or poisonous organisms. There are even aposematic sounds and odors.

MIMICRY

Different species sometimes resemble one another for reasons of defense. For instance, should a certain coloration or pattern or design be aposematic, many of several different species may evolve a similar aposematic coloration because predators learn to avoid anything looking like that. Such a phenomenon, where both the **model** and the **mimic** are unpalatable, is known as **Mullerian mimicry**. Interestingly, there are some species that are not advertising an unpalatable chemical or physical defense, even though they still resemble other aposematically colored species. These animals are called mimics because they mimic a model, and the phenomenon is termed **Batesian mimicry**.

CAMOUFLAGE

Many organisms blend in with their background. **Camouflage** is a means of disguise that conceals an organism. As humans, we often see camouflage in visual terms (animals colored or shaped to blend in with their background), largely because we are visually oriented. However, there are other means of disguise, such as those that make an animal difficult to hear, smell, or feel. Both prey and predators may be well camouflaged. Not only animals use camouflage, many plants do too. However, few biologists have yet to record such instances.

One type of camouflage is **countershading**. Instead of being the same color all over, an organism may be darker above and lighter below, as is the case with many fish. Therefore, when viewed from above, fish blend in with the water and, when viewed from below, they blend in with the bright sky.

Disruptive coloration breaks up an animal's silhouette so the organism doesn't stand out. Rather than being a uniform color, it may have blotches or patches that make it look like part of the vegetation, or whatever the organism is blending in with. **Cryptic coloration** hides an animal against its background by being the same color or pattern. For instance, an animal that lives on sand might be sand colored or one that lives in the trees might be green to blend in with the foliage.

Key Terms

abiotic
aliens
alpine tundra
aposematic coloration
atmospheric cycles
autotroph
bacteria
Batesian mimicry
behavior
behavioral ecology
biogeochemical cycles
biomass
biome
biopsychology
biotic potential
birth rate
blue-green bacteria (cyanobacteria)
boreal forest (taiga)
camouflage
carbon cycle
carnivore
carrying capacity
chaparral (temperate shrubland)
chemical defense
chemosynthesis
climate
climax community
clumped spacing
commensalism
community
competition
competitive exclusion principle
contest competition

continental shelf
coral reef
countershading
cryptic coloration
cyanobacteria (blue-green bacteria)
death rate
deciduous
decomposer
defense
denitrifying bacteria
density
density-dependent factors
density-independent factors
desert
dispersion
disruptive coloration
dominance
ecological homeostasis
ecology
ecosystem
ecotone
ectoparasite
endoparasite
estuary
ethology
extirpation
food chain
food web
freshwater wetlands
fundamental niche
fungi
generalists
gradient zone

greenhouse effect
gross primary productivity
habitat
herbivore
heterotroph
home range
humidity
interspecific
intraspecific
life tables
limnologists
littoral zone
logistic curve
longevity
marine communities
microclimate
mimicry
model mimic
Mullerian mimicry
mutualism
natural selection
nekton
net productivity
niche
nitrifying bacteria
nitrogen cycle
nitrogen-fixers
nonrandom spacing
omnivore
open ocean
parasitism
permafrost
phosphorus cycle
physical defense
plankton
ponds and lakes
population
population growth
predation
predator
primary consumer
primary productivity
producer
quaternary consumers
rainfall

random spacing
realized niche
recruitment
reproductive strategy
scramble competition
secondary consumer
secondary productivity
sedimentary cycle
social hierarchy
society
specialists
species diversity
species turnover
standing crop
streams and rivers
sublittoral zone
succession
succulent
survivorship curve
symbiosis
taiga (boreal forest)
temperate deciduous forest
temperate desert
temperate evergreen forest
temperate forest
temperate grassland
temperate rain forest
temperate shrubland (chaparral)
temperate woodland
temperature
territory
tertiary consumers
tidal marsh
timberline
transitional zone
trophic level
tropical rain forest
tropical savanna
tropical seasonal forest
tropical thornwood
tundra
uniform spacing
urban ecosystems
weather

Chapter 15 Self-Test

QUESTIONS TO THINK ABOUT

1. Compare and contrast the following terms: species, population, community, ecosystem.
2. What are the differences between r- and K-selected species? Give two examples of each, and provide two examples of intermediary species that are neither, but have characteristics of both.
3. Why does it work for some species to be generalists, while others are specialists? Compare the risks and the benefits.
4. Write a brief paragraph explaining four distinctly different scenarios in which four introduced species (imagined or real) have affected community dynamics in entirely different manners.
5. Describe the conditions that would render a forest a climax community. What might change this climax community so it is no longer a stable assemblage of species?
6. List ten different ecosystems and briefly describe them.

MULTIPLE-CHOICE QUESTIONS

1. Ultimately, the _____ is the energy source in most ecosystems.

   a. sun
   b. moon
   c. stars
   d. gravity
   e. air

2. Light energy is used by plants to convert, during photosynthesis, _____ _____ and _____ into carbohydrates.

   a. water and carbon dioxide
   b. bread and water
   c. oxygen and water
   d. fats and proteins
   e. oxygen and carbon dioxide

3. Organisms such as plants, that manufacture their own organic food molecules from inorganic substances, are the _____, and are often called _____.

   a. autotrophs, decomposers
   b. chemosynthetics, herbivores
   c. secondary consumers, tertiary consumers
   d. predators, carnivores
   e. producers, autotrophs

4. Organisms that acquire food from dead plants and animals are called
_____.

    a. herbivores
    b. heterotrophs
    c. decomposers
    d. chemosynthetics
    e. predators

5. Organisms that eat plants, especially those that eat only plants, are called
_____.

    a. predators
    b. carnivores
    c. autotrophs
    d. herbivores
    e. producers

6. Animals that eat both plants and animals are called _____.

    a. pyramid builders
    b. food webbers
    c. omnivores
    d. detritivores
    e. omnitrophs

7. The major terrestrial communities, those areas with distinct categories and
types of organisms, are often called _____.

    a. rain forests
    b. biomes
    c. ecoclines
    d. forests
    e. prairies

8. Characterized by more than 400 cm (over 150 inches) of rainfall and warm
temperatures during most of the year, rather than just during a specific
warm or rainy season, _____ are wet and green all year round.

    a. arctic tundra
    b. tropical oceans
    c. boreal forests
    d. tropical rainforests
    e. tropical seasonal forests

9. Which of the following has more species of deciduous trees than species of
trees that keep their leaves or needles all year long?

    a. tropical rain forests
    b. tropical seasonal forests
    c. temperate zone forests
    d. a and b
    e. b and c

10. Communities that have less than 20 cm (8 inches) of rain a year are known as _____.

   a. deserts
   b. rain forests
   c. grasslands
   d. all of the above
   e. none of the above

11. The newest, fastest growing ecosystems being developed by people where they move in large numbers, are known as _____.

   a. tropical rainforests
   b. salt marshes
   c. tundra
   d. streams and rivers
   e. urban ecosystems

12. Much of America's Midwest and the Soviet Union's Ukraine, which now support productive agriculture, were originally _____.

   a. tropical rainforests
   b. boreal forests
   c. deserts
   d. tundra
   e. temperate grasslands

13. The _____ forest, often called the taiga, stretches through much of Canada and Siberia. This biome is dominated by spruce trees, pines, and firs that do well under what would be harsh conditions for most other species.

   a. tundra
   b. boreal
   c. temperate
   d. tropical
   e. hardwood

14. Vegetation above the tree line primarily consists of _____.

   a. sedges and grasses
   b. mosses and lichens
   c. herbaceous flowering plants
   d. dwarf woody shrubs
   e. all of the above

15. The area between the high and low tide marks is known as the _____.

   a. littoral zone
   b. coral reef
   c. shore
   d. rocky intertidal zone
   e. beach

16. Along with estuaries and tropical rain forests, _____ are among the most productive ecosystems in the world.

    a. deserts
    b. tundras
    c. beaches
    d. coral reefs
    e. temperate shrublands

17. Specific biologically active minerals cycle through plants and animals and the environment; these cycles are known as _____.

    a. biogeochemical cycles
    b. pyramid of numbers
    c. pyramid of biomass
    d. biomes
    e. none of the above

18. The increase in average worldwide atmospheric temperatures, attributed to additional CO_2 in the atmosphere, is known as the _____.

    a. bicarbonate cycle
    b. carbon cycle
    c. nitrogen cycle
    d. greenhouse effect
    e. acid rain effect

19. Because amino acids and proteins need nitrogen, and because gaseous nitrogen is not the form that can be readily used by most organisms, it is through the conversion of gaseous nitrogen, by certain bacteria and blue-green bacteria, into an aqueous form, that this nitrogen becomes available to other organisms. This conversion is known as _____.

    a. nitrogen-fixing
    b. nitrogen-breaking
    c. nitrogen-mending
    d. alchemy
    e. composting

20. Because fixed nitrogen is so important to plants, it is a key ingredient in many fertilizers. However, because this fertilizer is so expensive, farmers often choose to rotate their crops, occasionally growing plants with symbiotic nitrogen-fixing bacteria living in nodules attached to the roots. Such plants are members of the _____ family.

    a. rose
    b. buttercup
    c. lily
    d. heath
    e. pea

21. The bacteria that live under anaerobic conditions (no oxygen), such as those living deep in the soil or in the mud, complete the nitrogen cycle by returning nitrogen back to the air. Such bacteria are known as _____.

    a. sedimentary bacteria
    b. atmospheric bacteria
    c. phosphorus bacteria
    d. nitrifying bacteria
    e. denitrifying bacteria

22. Plotting the fate of the young and their chances of survival at each age enables researchers to construct _____.

    a. survivorship curves
    b. increase curves
    c. biotic potential curves
    d. carrying capacity curves
    e. niche breadth curves

23. Since the biotic potential is higher for fruit flies than it is for people, flies have a higher _____.

    a. a
    b. d
    c. K
    d. r
    e. q

24. The number of individuals of a particular species that a specific environment can support indefinitely, often referred to by the letter K, is known as the _____.

    a. carrying capacity
    b. specific environment
    c. species diversity
    d. population size
    e. ecological homeostasis

25. Many species are being introduced to areas where they are not native; such exotic species are often referred to as _____.

    a. Greeks
    b. out-of-towners
    c. Martians
    d. aliens
    e. Batesians

26. When individuals directly compete for a resource without regard to other individuals that may also be using that resource, it is known as _____.

   a. contest competition
   b. scramble competition
   c. realized competition
   d. unrealized competition
   e. lack of competition

27. Most animals have an area where they spend the majority of their time. This includes where they may interact with other members of the same species, look for food, rest, and sleep. Such an area is called a _____.

   a. territory
   b. niche
   c. home range
   d. symbiosis
   e. habitat

28. Long-lasting arrangements of animals where individuals change over time due to births and deaths, but the continuity of the social structure remains intact, are known as _____.

   a. pecking orders
   b. dominance-subordinance relationships
   c. mutualistic relationships
   d. symbiotic relationships
   e. societies

29. An arrangement between species where each benefits from one another and neither is harmed by the relationship is known as _____.

   a. parasitism
   b. mutualism
   c. commensalism
   d. monogamy
   e. polygamy

30. Instead of standing out, many organisms blend in with their background. Such a means of disguise that conceals an organism is known as _____.

   a. camouflage
   b. disruptive coloration
   c. chemical defense
   d. physical defense
   e. aposematic coloration

ANSWERS

1. a	9. e	17. a	25. d
2. a	10. a	18. d	26. b
3. e	11. e	19. a	27. c
4. c	12. e	20. e	28. e
5. d	13. b	21. e	29. b
6. c	14. e	22. a	30. a
7. b	15. a	23. d	
8. d	16. d	24. a	

CHAPTER 16

Viruses and Subviruses

Viruses and **subviruses** are not cellular. Therefore, by definition they are not living because their primary units are not cells. Even so, they interact with living organisms in many fundamental ways, interfering with their cellular processes, and are capable of causing a number of very real symptoms. In fact, these tiny, inscrutable, nonliving entities are responsible for over 60 million common colds each year in America alone, as well as for a number of other major scourges and epidemics, including AIDS.

VIRUSES

The fact that viruses share several properties of most living organisms provides good reason to include them in their own noncellular category of life. Were this to be sanctioned by most scientists, it would require expanding the definition of life as it currently stands to include noncellular entities that contain genes. Should such a revision ever occur, viruses might form their own kingdom.

Viruses are, on the average, from 10 to 100 times smaller than the typical bacterium, making them too small to be seen by most optical microscopes. In 1931, the invention of the electron microscope broke this light barrier. And X-ray crystallography, a technique by which X-rays are diffracted through crystallized virus particles to reveal their molecular structure, enabled researchers to study these forms.

Like some obligate intracellular bacteria, viruses are parasitic and unable to reproduce without having cells to inhabit. Viruses, like living cells, contain nucleic acids, which are enclosed in a protective coat of protein, sometimes called the **viral capsid**, which ranges from 20 to 250 nanometers across (1 nanometer = 1 millionth of a millimeter).

Outside cells, viruses neither reproduce, feed, nor grow. And, unlike living cells, viruses do not metabolize; that is, they do not generate their own energy. Instead, with the information contained in their viral DNA or RNA, they overpower other cells, inserting their nucleic acids into their host's cell to direct the production of more viruses by utilizing the host's cellular machinery. While all other organisms contain both DNA and RNA, viruses contain only one or the other.

Outside of a host cell, a virus is inert, incapable of reproduction, or of any metabolic functions that would identify it as living. However, each of the many different types of viruses "identify" receptor sites on a potential host's outer coat of protein and thereby "know" which cells to attack. A virus may infect a host cell either by attaching to the host's protein coat while injecting the viral DNA or RNA into the host or by entering the host intact. Once inside, the viral capsid dissolves and the viral DNA or RNA acts as a template for the manufacture of viral components. That is, the virus attaches its genetic material to that of the host and "tricks" it into producing more viruses through the same mechanisms the cell normally uses to replicate itself. In time, the virus particles are assembled within the host cell. Then, by lysing (dissolving) the cell membrane, the new viruses leave the host and infect new, uninfected cells (see Figure 16.1).

Viruses' capacity to interfere with and inject viral genetic information into a host's cells may play an important and, possibly, even crucial role in evolution. By rearranging the DNA in chromosomes and by transferring genes from one species to another, viruses may be moving genetic material among plants and animals, sometimes imparting new characteristics that are adaptively significant. In fact, it is possible that certain viruses may have an evolutionarily beneficial effect over time.

It has been suggested that viruses are more closely related to their hosts than to one another, having perhaps originated as nucleic acids that escaped from cells and began replicating on their own, but returning to the cells for necessary chemicals and structures.

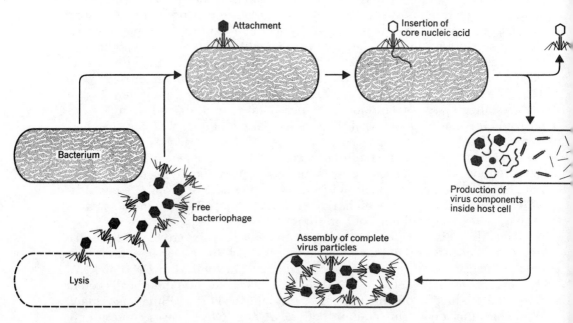

Figure 16.1 *Proliferative cycle of a virus; here the host cell is a bacterium and the virus is a bacteriophage.*

Some viruses known as **bacteriophages** attack only bacteria (see Figures 16.2 and 16.3). Others attack only eukaryotic cells. And many are extremely specific with regard to the type of cell they will attack. The major types of viruses that attack humans include cold viruses, or **rhinoviruses**, which cause most common colds.

The **influenza virus** is the fastest mutating virus known, capable of rapidly changing the outer protein coat through succeeding generations of the flu. People can therefore catch the flu more than once a year, since they have no antibodies to the new virus. RNA viruses cause measles, rubella (German measles), and mumps—all childhood diseases. Another childhood disease, called fifth disease, is caused by **parvovirus**.

Different forms of **herpes virus** include those that cause cold sores, genital sores, chicken pox (or, if it's reactivated, shingles), and Epstein-Barr virus, which causes mononucleosis.

Papillomaviruses, of which forty-six types are known, cause plantar warts, genital warts, and certain wartlike rashes. **Hepatitis** is also caused by a virus.

Retroviruses are a group of viruses named for their backward (retro) sequence of genetic replication as compared to other viruses. The **AIDS** viruses are in this group. Another well-known disease caused by a viral infection is rabies.

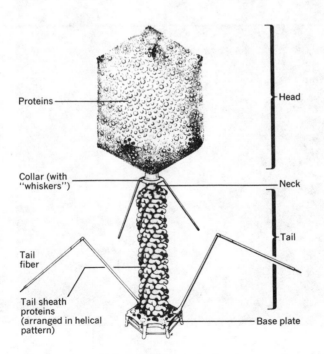

Figure 16.2 *Diagram of a virus that attacks bacteria, known as a bacteriophage (only two of the six collar whiskers and two of the six tall fibers are depicted here).*

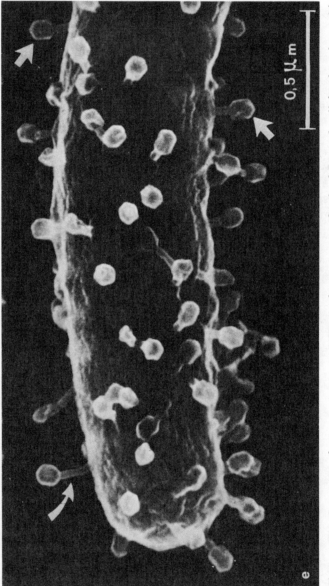

Figure 16.3 *Scanning electron microscope (SEM) photomicrograph of a bacterial cell (Escherichia coli) being infected by many bacteriophages.*

SUBVIRUSES

The smallest infectious agents known to researchers are termed **subviral infectious agents**, or **subviruses**. Scientists have identified at least six different strains: **satellite viruses, virinos, viroids, virusoids, virogenes**, and **prions**.

Members of one of the better understood strains, prions, range in size from considerably smaller than viruses, sometimes 100 times smaller, to almost as large as mitochondria and bacteria. Prions have been found to cause certain diseases and are implicated as the cause of others. Included in this list of diseases that prions seem to promote are scrapies and several similar degenerative brain diseases.

It has been theorized that prions may be radically different from any other known self-replicating entities. There is no evidence that prions contain any nucleic acids, DNA and/or RNA; instead, they appear to be little more than dots of protein. Even if they were found to contain nucleic acids, prions are so small that there is little chance they contain a nucleic acid any longer than 50 nucleotides. This is not large enough to encode a protein containing more than about 12 amino acids.

Despite indications to the contrary, it has even been suggested that prions may actually be conventional viruses, but this is quite unlikely. It appears equally unlikely that they will be found to represent a new category of proto-organismal material that reproduces in living cells, employing a technique that has yet to be elucidated. It has even been suggested that they may reproduce using a techique similar to that employed by viruses, without being viruses.

Some researchers have suggested that the mode of prion reproduction might involve fracture and continued growth, which would explain their small and uncertain molecular weights, their rod-like appearance, their varying lengths, and the unpredictability of which amino acid occurs terminally. The most recent work has shown that prions may be proteins produced somewhat abnormally by infected genes that somehow go awry.

Among the other subviruses are the viroids, minute rings of RNA that infect certain plants. Virusoids appear to be loops of RNA that occur inside regular viruses. Virinos, like viruses, need an outer coat of protein, which they are unable to make on their own, but which they induce host cells to manufacture. Virogenes are otherwise normal genes that generate infectious particles under certain circumstances. Satellite viruses are tiny pieces of RNA that make full-size viruses work for them. These tiny nucleic acids multiply inside viruses that are inside cells.

Key Terms

AIDS   hepatitis
bacteriophage   herpes virus

influenza virus	subviruses
papillomavirus	viral capsid
parvovirus	virinos
prions	virogens
retrovirus	viroids
rhinovirus	virus
satellite viruses	virusoids
subviral infectious agents	

Chapter 16 Self-Test

QUESTIONS TO THINK ABOUT

1. Describe viruses. Are they considered living? Support your answer.
2. What might account for the origin of viruses?
3. How do viruses increase in number? Describe the different mechanisms.
4. What are five diseases affecting humans that are caused by viruses?
5. Describe similarities and differences between viruses and subviruses.

MULTIPLE-CHOICE QUESTIONS

1. Viruses contain

    a. nucleic acids
    b. a protein coat
    c. DNA or RNA
    d. viral capsid
    e. all of the above

2. Viruses do not

    a. metabolize
    b. generate their own energy
    c. replicate (or duplicate, or reproduce) without injecting cells
    d. all of the above
    e. none of the above

3. The information contained in the viral DNA or RNA is

    a. inserted into its host's cellular machinery
    b. contained in the viral nucleic acids that are inserted into their host's DNA
    c. used to direct the host to produce more viruses
    d. all of the above
    e. none of the above

4. Each of the many different types of viruses "know" which cells to attack by identifying ——————— on the potential host's outer —————.

    a. receptor sites, protein coat
    b. nucleic acids, viral capsid
    c. receptacles, bacteriophages
    d. all of the above
    e. none of the above

5. Viruses that attack only bacteria are known as ———————.

    a. DNA viruses
    b. RNA viruses
    c. retroviruses
    d. bacteriophages
    e. all of the above

6. Some viruses infect a host cell by

    a. attaching to the host's protein coat while injecting the viral DNA or RNA into the host
    b. entering the host intact
    c. injecting viral genetic information into a host's cells
    d. all of the above
    e. none of the above

7. It is possible that viruses may be moving genetic material from

    a. plants to animals
    b. animals to plants
    c. plants to plants
    d. animals to animals
    e. all of the above

8. Viruses may prove, in some cases, to be the simplest of

    a. all symbionts
    b. all parasites
    c. all living things
    d. all of the above
    e. none of the above

9. Prions have been said to be

    a. the smallest infectious agents known
    b. the largest infectious agents known
    c. 100 times smaller than viruses to almost as large as mitochondria and bacteria
    d. the cause of certain diseases
    e. a, c, and d

10. Prions

    a. have been found to contain nucleic acids
    b. have not been found to contain nucleic acids
    c. are viruses
    d. are not viruses
    e. b and d

11. Recent work has shown that prions may be

    a. proteins produced somewhat abnormally by "infected" genes that some-
       how go awry.
    b. bacteria
    c. viruses
    d. protists
    e. mitochondria

ANSWERS

1. e	4. a	7. e	10. e
2. d	5. d	8. d	11. a
3. d	6. d	9. e	

CHAPTER 17

Monera

Unlike viruses and subviruses, which are not cellular, the members of the king-dom **Monera**, including **bacteria** and **blue-green bacteria** (sometimes called cyanobacteria, or blue-green algae), are composed of true cells. Monerans are all **prokaryotic**; that is, their cells lack most organelles, they do not have a membrane-bound nucleus, and most occur as single-celled organisms (see Figures 17.1 and 17.2).

Of the 15,000 described species, many exist as a series of cells occurring in long filaments or as more complex colonies. Scientists are discovering bacteria that form complex communities, hunt prey in groups, and secrete chemical trails for the directed movement of thousands of individual bacterial cells.

In comparison to most single-celled eukaryotes, individual bacterial cells are smaller and far more abundant, representing a remarkably important component of nearly all ecosystems. Without bacteria, life on earth could not exist as we know it. Bacteria represent some of the most important groups of **decomposers**; without them, dead organisms would not decay properly. Many nutrients would remain locked up in corpses forever. Geochemical recycling of the earth's nitrogen, carbon, and sulfur, which are critical to life, would not occur without bacteria. Chemicals such as nitrates, which certain plants use for protein synthesis, are produced by some species of bacteria. Certain bacteria are heterotrophic; that is, they procure their food by feeding on organic material formed by other organisms. Other species of bacteria are photosynthetic, capable of synthesizing their organic molecules from inorganic components, using the energy from the sun. One group of bacteria, the **mycoplasmas**, are the smallest known cells that grow and reproduce without needing a living host. Their diameters range from 0.12 to 0.25 micrometers.

Probably because of the small size of most types of bacteria, their rapid rate of cell division, remarkable metabolic versatility, and ability to live practically anywhere, they are the most numerous organisms on earth. Under optimal conditions, a population can double in size every 20 or 30 minutes. Species of bacteria are found thriving on icebergs, in hot springs, at the bottom of the oceans, in freshwater, on land, in the soil, and even in aviation fuel.

Although most bacteria use oxygen in their metabolic processes, there are many species that use alternative pathways, surviving perfectly well without any oxygen. Some species have the ability to form spores, which are inactive, thick-walled forms that survive for long periods without water or nutrients in what otherwise would be unfavorable conditions.

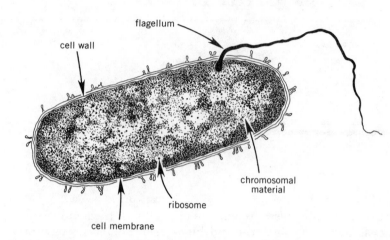

Figure 17.1 *Diagram of a bacterial cell, illustrating a prokaryotic cell. Most prokaryotes have few, if any membranous organelles within the cell, no nuclear membrane, no mitochondria, and no endoplasmic reticulum. Generally, there is a cell wall outside the plasma membrane. Flagelli, when present, are composed of protein, and do not have any microtubules.*

Bacteria were first discovered in 1676, but it was not possible to learn very much about them. In the nineteenth century, Louis Pasteur went as far as was possible without the aid of the subsequently developed electron microscope or advanced biochemical techniques, which enabled later researchers to study these small organisms in considerably more detail.

Being prokaryotes, bacteria have cells that differ from eukaryotes in the following ways.

1. *Cell walls.* Prokaryotic **cell walls** are composed of a polymer of glucose derivatives attached to amino acids. This substance is termed a **mucocomplex**. Some bacteria have an additional outer layer of a polymer composed of lipid and sugar monomers, which is termed **lipopolysaccharide**. Many bacteria can secrete polysaccharides that allow them to stick to things. Cell walls of **cyanobacteria** (blue-green bacteria) tend to be covered with gelatinous material.

2. *Plasma membrane.* Inside the cell wall of some bacteria is a **plasma membrane** that coils and loops, creating a unique structure known as the **mesosome**, which may be important in cell division.

3. *Other internal membranous structures.* Some prokaryotes have internal membranous structures containing photosynthetic pigments and related enzymes. Of the aerobic bacteria, those that use molecular oxygen for cellular respiration, some have internal membrane systems containing respiratory enzymes.

4. *Ribosomes.* The only organelles that consistently occur in prokaryotes are **ribosomes**, on which messenger RNA (mRNA) is found. The mRNA carries instructions from the genes to the ribosomes, where protein synthesis

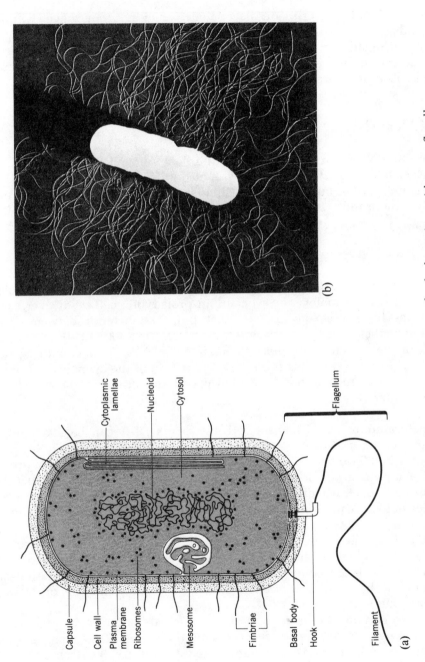

Figure 17.2 *(a) Composite bacterium; (b) an electron micrograph of a bacterium with many flagella.*

Capsule
Cell wall
Plasma membrane
Ribosomes
Mesosome
Fimbriae
Basal body
Hook
Filament
(a)

Cytoplasmic lamellae
Nucleoid
Cytosol
Flagellum

(b)

occurs. Prokaryotic ribosomes are smaller than those found in eukaryotic cells.

5. *Flagella*. Some bacteria are flagellated, meaning, they have whiplike appendages, extending singly or in tufts, that propel the cells. The **flagella** of higher organisms consist of a hollow cylinder containing nine pairs of fibrils surrounding two central fibrils. A bacterial flagellum consists of a single fibril of contractile protein.

PROKARYOTIC DNA

Prokaryotic DNA (deoxyribonucleic acid) differs from **eukaryotic DNA** in that it is associated with different proteins. It also differs from eukaryotic DNA in that it is not paired, but is circular. Circular DNA molecules consist of only about one thousandth the DNA found in eukaryotic cells.

REPRODUCTION

Most bacterial cells reproduce by the simple cell division, **binary fission**. Neither mitosis nor meiosis ever occurs in prokaryotic cells; however, some prokaryotes have a sexual process that transfers material between cells. Occasionally these bacterial cells will transfer DNA to another cell, after which some of the new DNA will replace the recipient's DNA. To date, nothing analogous to a sexual system has been observed in any of the cyanobacteria.

There are three methods by which genetic material may be transferred between bacteria.

1) **Transformation**. One bacterial cell breaks; its DNA can be taken up by another bacterial cell.
2) **Conjugation**. Two bacterial cells come together and are joined by a protein bridge, a **pilus**, through which DNA fragments pass from cell to cell.
3) **Transduction**. A bacteria-attacking virus, known as a **bacterial virus**, or **bacteriophage**, carries bacterial DNA from one bacterial cell to another.

Each of the three methods can result in the transfer of DNA fragments from one bacterial cell to another. During the transfer, sometimes homologous DNA fragments, those containing the same type of genetic information, are substituted in the recipient's circular DNA without a net increase or decrease in the total amount of circular DNA.

It is not certain how important **genetic recombination** is for prokaryotic evolution. However, despite the fact that **mutations** (inheritable changes in the organism's genetic material) occur infrequently, prokaryotes do have a high degree of genetic variability and therefore evolve quickly. When it exists, their rapid rate of evolution is usually attributed to their great numbers, and their incredible reproductive rate, as well as mutations and genetic recombinations. Knowledge of such DNA recombination led to research using viruses that transmit DNA fragments to other types of organisms. This research is expected to lead to human gene therapy.

SPORES

Many prokaryotes are also capable of producing a dormant stage known as a **spore**. Unlike the spores of other organisms, this is not a reproductive unit. Rather, bacterial spores function wholly as units that contain stored food and are highly resistant to desiccation, as well as to extremely hot and cold temperatures. Bacterial spores have been shown to survive temperatures as cold as $-252°C$, and some may be able to live for thousands of years. When conditions become favorable, the bacterial spore germinates into a new cell.

HETEROTROPHIC BACTERIA

Most bacteria selectively absorb organic molecules through the cell wall, rather than manufacturing all their organic nutrients internally, or autotrophically. These **heterotrophic bacteria**, along with fungi, are important decomposers because they secrete enzymes that digest large organic molecules into smaller molecules that can then be absorbed.

Species of bacteria have been found thriving in just about every habitat, including inside all animals. Many of these bacterial species do no harm, but some do cause disease. Others, bacterial **symbionts**, are vital to their hosts; some of these live in the gut of their hosts, digesting materials otherwise difficult to digest. In the case of termites, for example, their symbiotic bacteria digest cellulose into smaller molecular constituents that are then absorbed by the cells in the termite's gut. Both the bacteria and the termites benefit from such a relationship.

Bacteria that live inside cells are known as **endosymbionts**. Some researchers say that the distinction between many cellular organelles and their intracellular symbionts may be a function of when the association first took place and the degree to which the different elements have become interdependent.

Some bacteria are **pathogenic**, or capable of causing disease. Some do so by destroying cells, and others by producing **toxins**, chemicals that can harm the host. **Antibiotics**, substances produced by some bacteria and fungi, arrest the growth of or destroy the agents of specific infectious diseases. Some antibiotics have been particularly effective in controlling diseases caused by specific species of bacteria. However, since antibiotics were first discovered during World War II, rapid bacterial evolution has favored resistant strains, as in the case of certain strains of sexually transmitted diseases that can no longer be controlled with the antibiotics that previously were effective.

Some bacteria have an **episome**, which is a DNA segment not attached to the circular DNA. The episome can be transmitted from one individual to another of the same or even different species. Sometimes the genes involved in drug resistance are located on the episome and are capable of being rapidly transmitted throughout the bacterial population within a relatively brief period after a new drug reaches the market. Some of the most widely known bacterial diseases are syphilis, gonorrhea, botulism, bubonic plague, diptheria, and tetanus.

Without going into all the different categories of bacteria, it should be said that many bacteriologists classify them into two major subdivisions: the **Archaebacteria** and the **Eubacteria**. The Archaebacteria are thought by some to represent the oldest group of organisms still living. They are distinctive with regard to their biochemical characteristics. Their membranes have an unusual lipid composition, their transfer RNAs (tRNA) and RNA polymerases are distinctive, and their cell walls do not contain peptidoglycan, which is found in all the Eubacteria. The two major groups of Archaebacteria are the **Methanogens** and the **Thermoacidophiles**.

The Eubacteria represent a large assemblage of species that reproduce by **binary fission**, the process whereby one cell divides asexually into two daughter cells. Eubacteria are often described in terms of their shape. Those that are rod-shaped are called **bacilli**, spherical Eubacteria are known as **cocci**, and spiral Eubacteria are **spirilla**. Some bacteria are gram-negative bacteria and others gram-positive; these terms merely describe whether the bacteria in question retain a violet dye used in Gram's staining technique.

There are many other major bacterial groups, including the cyanobacteria (blue-green bacteria), as well as the purple, brown, and green sulfur bacteria, sometimes called the pseudomonadales, spirochaetes, actinomycetes, rickettsias, and mycoplasmas represent other important groups.

CHEMOSYNTHETIC BACTERIA

Bacteria are remarkably diverse with regard to their metabolic pathways. Some are **aerobic**. Others do not require molecular oxygen in their breakdown of food to release energy; these forms are termed **anaerobic**. Most cyanobacteria have elaborate internal membranes containing photosynthetic pigments that synthesize organic compounds from inorganic materials using light energy. Fossils closely resembling living cyanobacteria have been found that indicate oxygen-producing photosynthesis existed more than 3.3 billion years ago.

Other types of bacteria have the capacity to synthesize high-energy compounds from inorganic materials without needing any light energy. These bacteria trap the energy released when oxidizing inorganic compounds. This form of autotrophic nutrition, where organic nutrients are manufactured from inorganic raw materials, involves the oxidation of various nitrogen and sulfur compounds. Even iron and molecular hydrogen are involved in certain chemosynthetic pathways. A few of the more common reactions are discussed below.

Energy is released when ammonia (NH_4^+) is oxidized into nitrite (NO_2^-) by adding oxygen.

$$2NH_4^+ + 3O_2 \rightarrow 2NO_2^- + 2H_2O + 4H^+ + energy$$

An alternative chemosynthetic process, employed by other bacteria, creates energy through oxidation when oxygen is added to nitrite (NO_2^-), synthesizing nitrate (NO_3^-).

$$2NO_2^- + O_2 \rightarrow 2NO_3 + \text{energy}$$

Another such process oxidizes sulfur (S) to sulfate (SO_4^{--}).

$$2S + 3O_2 + 2H_2O \rightarrow 2SO_4^{--} + 4H^+ + \text{energy}$$

NITROGEN FIXATION

Nitrogen is an important element in many molecules and in many chemical reactions. Although 78 percent of the atmosphere is gaseous nitrogen (N_2), it occurs in a very unreactive form. Before nitrogen becomes useful to any organisms, N_2 must first be broken into two atoms. This is done by organisms called **nitrogen fixers**, most of which are prokaryotic; many live in close association with certain plants. Some of these bacteria live in the root nodules of such plants as the legumes. Their role in the nitrogen cycle is described in Chapter 15.

Key Terms

aerobic
anaerobic
antibiotic
Archaebacteria
bacilli
bacteria
bacterial virus
bacteriophage
binary fission
blue-green bacteria
cell walls
chemosynthesis
cocci
conjugation
cyanobacteria
decomposers
denitrifiers
endosymbionts
episome
Eubacteria
eukaryotic DNA
flagella
genetic recombination

heterotrophic bacteria
lipopolysaccharide
mesosome
Methanogens
Monera
mucocomplex
mutation
mycoplasmas
nitrogen-fixing bacteria
pathogen
pilus
plasma membrane
prokaryote
prokaryotic DNA
ribosomes
spirilla
spore
symbiont
Thermoacidophiles
toxin
transduction
transformation

Chapter 17 Self-Test

QUESTIONS TO THINK ABOUT

1. Give several examples of Monera, and describe the characteristics that all Monera share.
2. What role do bacteria play in most ecosystems?
3. Describe different modes of bacterial reproduction.
4. How do heterotrophic and chemosynthetic bacteria differ?
5. What is nitrogen fixation, why is it important, and how does it occur?

MULTIPLE-CHOICE QUESTIONS

1. Genetic material may be transferred between bacteria by ——————— .
   a. transformation
   b. conjugation
   c. transduction
   d. none of the above
   e. all of the above

2. Inside the prokaryotic cell wall is a plasma membrane, which in some bacteria coils and loops, creating a unique structure known as the ——————— , which may be important in cell division.
   a. aster
   b. centromere
   c. centriole
   d. mesosome
   e. ribosome

3. The only organelles consistently found in prokaryotes are ——— ——— .
   a. ribosomes
   b. autosomes
   c. nuclei
   d. chloroplasts
   e. coenocytes

4. Many prokaryotes are capable of producing a dormant stage, known as a(n) ——————— , that contains stored food and is highly resistant to desiccation as well as to hot and cold temperatures.
   a. seed
   b. cyst
   c. pollen granule
   d. spore
   e. egg

5. _____ are substances derived from either fungi or bacteria that arrest the growth of or destroy the agents of specific infectious diseases.

   a. antibiotics
   b. enzymes
   c. lysosomes
   d. fungicides
   e. all of the above

6. Some bacteria have a(n) _____ , which is a DNA segment that is not attached to the circular DNA.

   a. mesosome
   b. ribosome
   c. episome
   d. lysosome
   e. chromosome

7. The process where one cell divides asexually into two daughter cells is known as _____ .

   a. mitosis
   b. meiosis
   c. transformation
   d. transduction
   e. binary fission

8. Spherical Eubacteria are known as _____ .

   a. episomes
   b. centromeres
   c. bacilli
   d. cocci
   e. spirilla

ANSWERS

1. e	3. a	5. a	7. e
2. d	4. d	6. c	8. d

CHAPTER 18

Protista

Protista constitutes a diverse kingdom containing thousands of species of single-celled organisms. Because many questions still persist concerning the ancestry of these organisms, deciphering which organisms should be classified in this kingdom is often a more arbitrary decision than most biologists would like. Because Protista is presented in the majority of biology texts as one of the five kingdoms, that's how the organisms are presented here.

Because multicellularity evolved many times, many multicellular organisms are more closely related to their ancestral unicellular lineages than they are to other multicellular organisms. This accounts for the reason that some members of the plant kingdom (such as the large multicellular algae) are sometimes considered to be multicellular protists. And certain members of the fungal kingdom (such as slime molds) are sometimes considered closer to the protistal lineage than to that of the fungi, and therefore are placed in the former. And in some classifications certain single-celled, heterotrophic protists are grouped with the animal kingdom. These are even respectable classifications in which all the major groups considered to be protists are placed in other kingdoms, and Protista is entirely dispensed with. The kingdom Protista as presented here, however, reflects the most widely accepted classification found in the majority of biology texts.

Discrepancies between different classifications are partially attributable to the way protists are defined. Rather than being grouped together by their shared characteristics, they are grouped by exclusion. That is, in addition to usually being unicellular, all protists are eukaryotes, so they are not included among the phylum Monera; since none develop from an embryo, they are not included among the phylum Plantae; since most do not develop from spores, they are not included among the phylum Fungi; and since none develop from a blastula, they are not included among the phylum Animalia. The organisms that remain tend to be those placed in this kingdom, Protista.

This kingdom includes the most simple, and often the most primitive, eukaryotic microorganisms and all their immediate descendants. Each protist cell has a nucleus and all the other eukaryotic properties. Members of this kingdom vary considerably in structure and physiology, ranging from heterotrophs (usually free-living, although there are parasitic forms) to photosynthetic autotrophs.

Protists appear to have evolved from a moneran type of ancestor. Protists possess specialized features such as endoplasmic reticulum, Golgi bodies, centrioles, chloroplasts, and mitochondria, as well as different kinds of vacuoles, granuoles, and fibrils. In addition, the average unicellular protist is considerably larger than the average moneran, and its cell division has become distinct from moneran cell division, having evolved mitotic and meiotic cell division.

It is theorized that the primitive protists were both plant-like and animal-like, having the capacity to obtain food by different mechanisms, as well as being able to photosynthesize additional food internally. There is considerable evidence that symbiotic relationships with prokaryotes living inside some early eukaryotes led to the development of chloroplasts and mitochondria. These organelles are contained in protists, as well as in many other more advanced eukaryotes.

Protista consists of several widely divergent phyla. Some of the unicellular, nonphotosynthetic protists are grouped as the **Protozoa**. These are subdivided into four classes: the **Mastigophora** (flagellates), the **Sarcodina** (amoebas), the **Ciliophora** (ciliates), and the **Sporozoa** (spore formers). The first three classes are identified according to their locomotor structures; however, sporozoans have no locomotor organelles, and instead they are characterized by their spores. To date, about 50,000 species of protozoans have been described.

Together, there are several other phyla that are often called true algae. They include about 25,000 described species, some of which belong to evolutionary lineages that were already well-developed more than 450 million years ago. Nearly all the members of these phyla are photosynthetic. They include forms that occur either as single cells, as filaments of cells, as plates or in planes of cells, or as a solid body. They range in size from unicellular microscopic organisms to giant multicellular forms such as the kelps, which often reach lengths exceeding 150 feet. A brief description of these photosynthetic groups is presented below.

There are three phyla of unicellular algae. **Euglenophyta (euglenoids)** live in fresh water, move by means of one to three flagella per cell, and have no cell wall. **Chrysophyta** usually include the **yellow-brown algae, yellow-green algae**, and the **diatoms**. They are mostly marine and contain pectic compounds, with siliceous materials providing the cell wall components. **Pyrrophyta (dinoflagellates)** live in marine environments, in fresh water, and in moist soil. They are characterized by having two flagella that beat in different planes, causing the organisms to spin. They often have distinctive, if not bizarrely shaped, cellulose walls. Like diatoms, the dinoflagellates are major components of the phytoplankton; they are aquatic, free-floating, photosynthetic, and usually microscopic.

There is another group of algae, sometimes called the true algae, discussed in Chapter 20. In a recent classification by Margulis and Schwartz, the true algae are grouped with the protists, along with the water molds, slime molds, and slime nets, forming a kingdom they call **Protoctista**.

Key Terms

Chrysophyta
Ciliophora
diatoms
dinoflagellates
euglenoids
Euglenophyta
Mastigophora
Protista

Protoctista
Protozoa
Pyrrophyta
Sarcodina
Sporozoa
yellow-brown algae
yellow-green algae

Chapter 18 Self-Test

MULTIPLE-CHOICE QUESTIONS

1. All protists

    a. together represent a kingdom
    b. are eukaryotic organisms
    c. are unicellular
    d. are multicellular
    e. a and b

2. All protists

    a. photosynthesize
    b. are heterotrophs
    c. have tissue differentiation
    d. all of the above
    e. none of the above

3. The following are protists:

    a. blue-green bacteria
    b. bacteria
    c. bryophytes
    d. euglenoids
    e. lycopsids

4. Unlike monerans, protists possess

    a. endoplasmic reticulum
    b. mitochondria
    c. Golgi bodies
    d. chloroplasts
    e. all of the above

5. The nonphotosynthetic protists are known as

   a. thallophytes
   b. bryophytes
   c. chlorophytes
   d. protozoa
   e. Chrysophyta

6. All of the following are protozoans, except for

   a. Sarcodina (amoebas)
   b. Ciliophora (ciliates)
   c. Mastigophora (flagellates)
   d. Sporozoa
   e. Tracheophyta

7. Which of the following is prokaryotic?

   a. dinoflagellates
   b. blue-green bacteria
   c. brown algae
   d. red algae
   e. diatoms

ANSWERS

1. e	3. d	5. d	7. b
2. e	4. e	6. e	

CHAPTER 19

Fungi

There is some evidence that the organisms classified as fungi arose from protists along several different evolutionary lines. In fact, depending on the classification, fungi are sometimes placed within the kindgom Protista, or within the kingdom Plantae, or in their own kingdom, **Fungi**.

Fungi are eukaryotic organisms; most exist in multicellular form, although some go through an amoeba-like stage, and others, such as yeast, exist in a unicellular form. Unlike the photosynthetic algae and plants, fungi do not photosynthesize but absorb food through their cell walls and plasma membranes.

The **slime molds** are different from most other fungi in that they are mobile during part of their life history. Some slime molds exist as a plasmodium, which is a **multinuclear (coenocytic)** mass of cytoplasm lacking cell walls. The **plasmodium** moves about and feeds in an amoeboid manner. The amoeboid mass is a slime mold's diploid phase.

Other slime molds have separate feeding amoebas that occasionally congregate into a **pseudoplasmodium** that then sprouts asexual fruiting bodies. Because they pass through an amoeba-like stage, slime molds are occasionally classified as protists. Slime molds are usually found growing on such decaying organic matter as rotting logs, leaf litter, or damp soil, where these viscous, glistening masses of slime are usually white or creamy in appearance, though some are yellow or red.

During its vegetative phase, the slime mold plasmodium moves about slowly, phagocytically feeding on organic material. Under certain conditions, the plasmodium stops moving and grows **fruiting bodies**, from which **spores** are released that upon germination produce **flagellated gametes**. The gametes fuse, forming zygotes that lose their flagella and become amoeboid. The diploid nucleus continually undergoes mitotic divisions without any cytokinesis, and the organism develops into a multinuclear plasmodium that usually reaches a total length of five to eight centimeters.

Most fungi secrete digestive enzymes that hydrolyze nearby organic matter into minerals and compounds that can then be absorbed. Chemicals that don't get absorbed, as well as the fungal waste products, enrich the surrounding area and become available to plants and other nearby organisms.

Fungi obtain their nutrition in any of three ways, or in any combination of these three ways: as **saprophytes**, living on dead organic matter; as **parasites**, attacking living plants or animals; and in **mycorrhizal associations**, in which they have a symbiotic relationship with plants, usually tree or shrubs.

Fungal spores are tiny haploid cells that float through the air, dispersing the fungi to new habitats. They are relatively resistant to high and low temperatures as well as to desiccation, and can survive long periods in an unsuitable habitat. When conditions become right, however, the spores germinate and grow. They absorb food through long, threadlike **hyphae**. The mass of branching hyphae creates the body of the fungus, called the **mycelium**. Mycelia grow, spreading throughout their food source. Some hyphae are **coenocytic**, having many nuclei within the cytoplasm. Others are divided by **septa** into compartments containing one or more nuclei. The rigid cell walls of the hyphae and fruiting bodies are composed of cellulose, or other polysaccharides, although some are composed of chitin.

The mycelium constitutes the largest part of the fungal body, yet few ever see mycelia because they are usually hidden within the source of food they are eating. Sometimes, however, they can be seen on the forest floor spreading over moist logs and dead leaves. When mycelia break into fragments, fungi can reproduce vegetatively. Each fragment may grow into a new individual fungus. Other methods of fungal reproduction involve the production of spores, which can be formed asexually or sexually. The spores are usually produced on structures that extend above the food source, where they can be blown away and travel to new environments. Slime molds send up spore-bearing fruiting bodies. The mycelia of mildew send up aerial hyphae that form spores. The fruiting bodies that most people are familiar with, those associated with such fungi as mushrooms, are huge compared to the tiny fruiting bodies that cover moldy bread and cheese.

Most fungi are either parasitic or symbiotic. Parasitism occurs when one individual benefits while the other is harmed, and symbiosis is a mutually beneficial relationship between two individuals. By far the majority of fungal species are terrestrial and reproduce both sexually and asexually. Many have mycelia that grow in a close, intimate manner with plant roots. In such a relationship, the plant benefits by receiving nitrogen and phosphorus, while the fungus benefits by receiving nutritious carbohydrates. The remainder of this chapter discusses some of the major groups of fungi.

OOMYCOTA: WATER MOLDS

The **water molds** and their relatives include the molds that grow on dead animals in the water. The powdery mildew found growing on Concord grapes is also a member of this group.

ZYGOMYCOTA

Zygomycota represent a group of fungi that, like the Oomycetes, have coeno-cytic hyphae. They also have chitinous cell walls. Although there are hundreds of species in this group, few people recognize any of them.

ASCOMYCOTA: SAC FUNGI

The **Ascomycota**, or **sac fungi**, form another group of fungi that is widespread although just a few species are familiar. Among this group's 30,000 or so species are the yeasts, certain bread molds, and the fungi that produce peni-cillin, as well as the species involved in making Roquefort and Camembert cheeses.

The yeasts are unicellular, but the Ascomycetes also include many multi-cellular types that form hyphae with perforated septa, allowing the cytoplasm and organelles such as ribosomes, mitochondria, and nuclei to flow from one cell to another (see Figure 19.1).

Asexual reproduction is common among the Ascomycetes. It occurs when-ever the projections known as **conidia** form and the asexual **conidiospores** pinch off. The sexual part of the life cycle involves two hyphae growing together so that the two nuclei become housed within the same cell. When these cells, called **dikaryons**, develop into the fruiting bodies known as **asci**, which are characteristic of the Ascomycetes, the two nuclei fuse inside each ascus (singular of asci). This is the process of fertilization. Then the diploid nucleus undergoes meiosis, forming four haploid nuclei. These undergo mito-sis, forming eight haploid nuclei that become the **ascospores**. When the ascus ruptures, the ascospores are liberated.

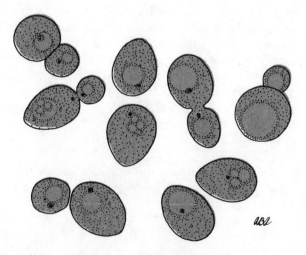

Figure 19.1 *Yeast cells in various stages of budding.*

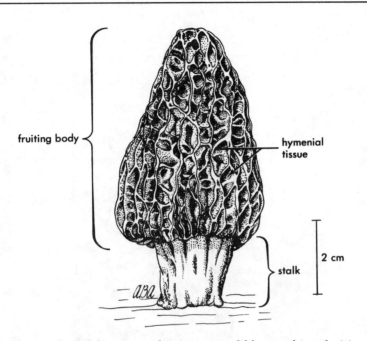

Figure 19.2 The mushroom known as the common edible morel is a fruiting body of the mass of branching hyphae, called the mycelium, growing underground.

BASIDIOMYCOTA: CLUB FUNGI

The **Basidiomycota,** or **club fungi**, include most of the common **mushrooms**. Their fruiting bodies are known as mushrooms (see Figure 19.2), **basidia**, or **clubs**; they are formed when two hyphae fuse. This is fertilization. A diploid nucleus is formed that undergoes meiosis, forming four haploid nuclei that move along thin extensions created by outgrowths of the cell walls. These nuclei are pushed to the edge of the club, where these **basidiospores** (spores) easily break off from their delicate stalks and are carried away by the slightest breeze. If they land in a suitable location, the spores germinate and grow hyphae, which form a mycelium that eventually sends up more fruiting bodies (see Figure 19.3).

IMPERFECT FUNGI

The **imperfect fungi** represent about 25,000 fungal species for which sexual reproduction has either been lost or has yet to be observed. Without information about their sexual stages, it has not been possible to identify the characteristic structures that would help specialists classify them appropriately. Accordingly, they have all been lumped together and called imperfect. Members of this group are responsible for ringworm and athlete's foot; both are fungi that infect people without ever sprouting fruiting bodies.

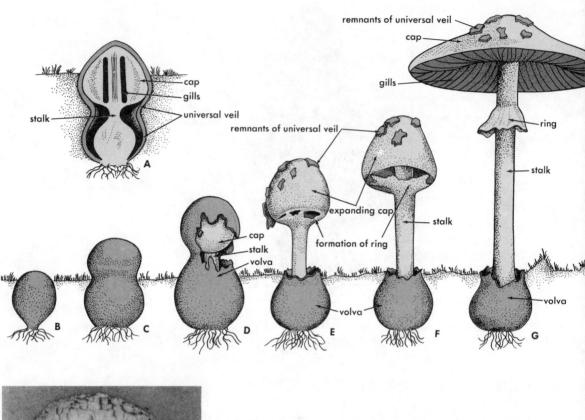

Figure 19.3 *Growth of a common poisonous mushroom (*Amanita*).*

MYCORRHIZAE

Mycorrhizal associations occur when the hyphae of a fungus grow around, between, and sometimes even into living plant root cells. Such associations have been found to occur in at least 90 percent of all the different plant families. Eighty percent of all the angiosperms (flowering plants) may have such associations. These relationships are symbiotic.

Plants benefit because the mycorrhizae mobilize nutrients by secreting enzymes that help to decompose the litter in the soil. And then, by acting as root hairs, they help to absorb the nutrients, especially nitrogen and phosphorus, by moving these nutrients from the soil into the root tissue. Mycorrhizae also secrete antibiotics that help reduce the plant's susceptibility to infection by pathogens. The mycorrhizae benefit by absorbing the chemicals and carbohydrates that constantly leak through the roots.

Many of the mushrooms seen under trees and shrubs are the fruiting bodies of the fungi that have a mycorrhizal relationship with the roots of the neighboring plants. One often sees certain species of mushrooms associated with certain species of plants because mycorrhizal relationships are often quite specific.

LICHENS

Lichens are symbiotic combinations of organisms living together intimately. The species involved are always a fungus and either a chlorophyte (green algae) or a cyanobacteria (blue-green bacteria). The fungi are always either members of Ascomycetes or Basidiomycetes. Although the fungi involved in lichens are usually not found growing alone, the photosynthetic portion of the lichens sometimes does live on its own. It is clear that the fungus living in a lichen benefits from the organic compounds obtained from the photosynthesizing member of the association. The algae may obtain water and minerals from the fungus, but this part of the interaction isn't well understood.

Because lichens are so tolerant of drought, heat, and cold temperatures, they are often the most important autotrophs found on recent lava flows, as well as on the stones used to construct buildings and gravestones. Lichens are also associated with dry, exposed soils, such as those in some deserts, and they also commonly occur in cold, exposed regions.

Most lichens reproduce either by **fragmentation**, when pieces break off and are blown elsewhere, or by spores produced by the fungal part of the lichen. The spores are blown or washed elsewhere, where they may grow and come in contact with an appropriate algal species. This marks the beginning of another lichen.

Key Terms

asci
Ascomycota
ascospores
basidia
Basidiomycota
basidiospores
club fungi
clubs
coenocytic
conidia
conidiospores
dikaryons
flagellated gametes
fragmentation
fruiting bodies
fungi

hyphae
imperfect fungi
lichen
multinuclear
mushrooms
mycelium
mycorrhizal associations
Oomycota
plasmodium
pseudoplasmodium
sac fungi
septa
slime mold
spores
water mold
Zygomycota

Chapter 19 Self-Test

QUESTIONS TO THINK ABOUT

1. Define the shared characteristics of organisms in the kingdom Fungi, and contrast this kingdom with the others.
2. List four groups of fungi and explain their differences, and yet the similarities that make them fungi rather than members of another kingdom.
3. What are lichens? Are they fungi? Why?
4. What are mycorrhizal associations? Who benefits from them? How and why?
5. Are all fungi always multicellular? If not, when aren't they? Give specific examples.
6. What part does a mushroom play in the life history of which fungi?
7. Athlete's foot is caused by a fungus, yet one never sees any fruiting bodies. Why?

MULTIPLE-CHOICE QUESTIONS

1. Slime molds have a diploid, coenocytic, amoeboid mass that is known as a
   _____.

   a. true fungus
   b. mushroom
   c. ascus
   d. plasmodium
   e. water mold

2. Fungi are never

   a. unicellular
   b. multicellular
   c. eukaryotic
   d. prokaryotic
   e. heterotrophic

3. Fungi

   a. photosynthesize
   b. absorb food through their hyphae
   c. have leaves
   d. have true roots
   e. contain chloroplasts

4. Together, the mass of branching hyphae create the body of the fungus, called
   a _____.

   a. plasmodium
   b. mycelium
   c. coenocyte
   d. slime mold
   e. lichen

5. Some hyphae are coenocytic, having many nuclei within the cytoplasm,
   and others are divided by _____ into compartments containing
   one or more nuclei.

   a. lichens
   b. mycelia
   c. cyanobacteria
   d. septa
   e. none of the above

6. Slime molds send up _____.

   a. mushrooms
   b. spore-bearing fruiting bodies
   c. seed bearing structures
   d. leafy parts
   e. none of the above

7. Many true fungi have mycelia that grow in a close, intimate manner with plant roots, where the plants benefit by receiving _____ and _____ while the fungus benefits by receiving nutritious _____.

   a. carbohydrates, nitrogen, phosphorus
   b. nitrogen, carbohydrates, phosphorus
   c. nitrogen, phosphorus, carbohydrates
   d. all of the above
   e. none of the above

8. Lichens involve the close association of a _____ and a _____.

   a. fungus, chlorophyte
   b. fungus, green algae
   c. cyanobacteria, fungus
   d. blue-green bacteria, fungus
   e. all of the above

9. When the hyphae of a fungus grow around, sometimes in between, and even within living plant root cells, the association is _____.

   a. mycorrhizal
   b. beneficial to the hyphae
   c. beneficial to the plant
   d. all of the above
   e. none of the above

ANSWERS

1. d	4. b	6. b	8. e
2. d	5. d	7. c	9. d
3. b			

CHAPTER 20

Plant Kingdom: Plantae

In most classification systems the plant kingdom **Plantae** includes several groups of simple photosynthetic organisms, sometimes known as eukaryotic algae, a reference to when people still grouped the prokaryotic blue-green bacteria with the algae. Euglenophyta, Chrysophyta, and Pyrrophyta, three phyla of unicellular eukaryotic algae often called plants, were described in Chapter 18. The term true algae is no longer a technical word since there are so many algal groups that are not directly related to one another. Nevertheless, the common terminology is presented in conjunction with the formal classification because it continues to be used in most texts. Of the following eukaryotic algae, **Chlorophyta (green algae)** are thought to be the ancestors of most modern plants because they contain the photosynthetic pigments chlorophyll a and b, as well as beta-carotene; they store their food reserves as starch; and their cell walls are composed of cellulose. They are mostly freshwater organisms, though some are marine.

Phaeophyta (brown algae) are almost all marine and are common in cooler oceanic regions. **Rhodophyta (red algae)** are mostly warm-water, marine species, though some are freshwater. Most of the common species of seaweed are members of the brown and red algal groups.

Each of the three preceding phyla vary according to: 1) types of photosynthetic they contain, 2) type of food reserves stored internally, and 3) components found in their cell walls.

These algae are not included in the plant kingdom merely because they photosynthesize, since many protists and monerans also have that capacity. Nor is tissue differentiation always a key factor in determining where to draw the line. Multicellular algae have no true roots, stems, or leaves. Their simple body form is termed a **thallus**, which is why they are sometimes called **thallophytes**. Many biologists call algae plants, but only lower plants.

The reason such matters are complex is that these organisms represent many different lineages and many different steps in a continuum connecting the most primitive forms of life, such as bacteria, to the most complex multicellular organisms. It isn't always clear just where one should draw the somewhat arbitrary lines that artificially separate each of the five described kingdoms—Monera, Protista, Fungi, Plantae, and Animalia.

A reason the algae have been kept from the plant kingdom in some classifications is their reproductive structures. Higher plants have reproductive structures encased inside a protective wall of sterile cells that protect the developing zygotes before they are released from the female reproductive organs, where they were produced. Algae lack a protective wall of nondividing (sterile) cells, and their zygotes do not develop into embryos until after having been released from the female reproductive organs.

When lower plants and higher plants are placed in separate kingdoms, then the higher plants are sometimes called the **Embryophyta** since the female reproductive organs retain the zygotes until after they have developed into embryos. In addition to the lower plants, the major groupings in the plant kingdom include the **Bryophyta** (mosses, liverworts, and hornworts) and the **Tracheophyta**, which include all the **vascular plants** (**psilopsids, club mosses, horsetails**, and **ferns**) and **seed plants** (**gymnosperms** and **angiosperms**). However, the name Tracheophyta, which means tube plant, has since been eliminated from the classification.

MOVE TO LAND

True algae are mostly restricted to aquatic environments. The move to land was accompanied by many adaptations to what would otherwise have been a hostile environment. Out of water, plants were met with dry conditions, ultraviolet light, and nutritional problems. In addition, in the air these plants no longer benefited from the surrounding water's buoyancy, so some structural support became necessary.

Terrestrial plants had to obtain water by a new means, since they were no longer bathed in it. The water had to be both procured and then moved from its point of uptake to the other structures. In addition, the photosynthetic products had to be transported from their specialized photosynthetic structures to the other parts of the plant. Vascular tissues evolved that solved all these problems.

Excessive water loss from evaporation had to be curtailed, while the moist tissues necessary to allow gaseous exchange for metabolic and photosynthetic purposes had to be maintained. This led to highly evolved mechanisms that controlled overall water loss, while enabling structures to remain moist.

Reproductive needs also had to be modified in terrestrial environments, where flagellated sperm cells no longer had the surrounding water in which to swim. And special structures evolved that protected the early stages of embryonic development from desiccation.

The modifications that differentiate many terrestrial plants from their aquatic counterparts are understood only in terms of the factors affecting plants living in terrestrial versus aquatic environments. The following mechanisms have helped make it possible for embryophyte plants to inhabit terrestrial habitats:

1. A waxy cuticle usually covers the aerial parts of the embryophyte plants, acting as waterproofing and preventing excessive water loss.
2. All embryophytes are **oogamous**. That is, they have two types of gametes, one of which, the female, is typically the large, non-motile egg cell, the **oogamete**.
3. All embryophytes have multicellular sex organs covered with a layer of protective cells that are sterile. The male sex organs are known as **antheridia**, and the female are **archegonia**. Within the sex organs the gametes are protected from desiccation.
4. All embryophytes have egg cells (oogametes) that are fertilized within the archegonia.
5. While inside the archegonium, the zygotes develop into multicellular diploid embryos that obtain some of their water and nutrients from the parent plant.
6. In addition to producing gametes, embryophytes produce **spores**, reproductive cells that develop directly into full-grown plants without first having to undergo fertilization by joining with another cell. The embryophyte **sporangia** produce spores that are covered with a protective jacket of sterile cells.

BRYOPHYTA: MOSSES, LIVERWORTS, AND HORNWORTS

The **bryophytes** represent about 25,000 species of mosses, liverworts, and hornworts. On the basis of any one characteristic it is difficult to distinguish them from the thallophytes, or from what are sometimes referred to as the true algae. In contrast to algae, which tend to be composed of either single cells, or filaments (sheets of cells), which can intertwine to create more complex body structures, bryophytes are rarely filamentous, except during one stage in the life history of the mosses. Instead, bryophytes are composed of cells that form tissues called parenchyma; they are characterized by loosely fitting cells that have thin walls of cellulose. In between the cells are intercellular spaces incorporated within the cellular network.

In bryophytes, the principal photosynthetic pigments are chlorophyll a and b. These are biochemically similar to the pigments of the chlorophytes, from which bryophytes probably arose. Their energy reserves are stored in the form of starch, and their cell walls contain cellulose.

Bryophytes are usually terrestrial. But they remain somewhat dependent on their ancestral aquatic environment. This has kept their distribution limited to moist environments or to environments that are moist during a critical period each year. These small plants need water for their flagellated sperm cells to swim from the antheridia to the egg cells in the archegonia. Without any vascular tissue, their ability to move fluids internally across long distances is limited. And since xylem, the vascular tissue in higher plants, is necessary for support, the upper size limit of these plants without such supportive tissue is kept at a minimum.

All the bryophytes have an **alternation of generations**, with a **sporophyte generation** (diploid) and a **gametophyte generation** (haploid). Among the larger, more complex algae, most of which have an alternation of generations, there is some tendency toward a reduction of the gametophyte, multicellular stage, and an emphasis on the sporophyte, multicellular stage. In both the brown and red algae, the sporophyte generation is prominent, as is the case in the vascular plants. This tendency is not apparent in the bryophytes, in which the haploid gametophyte stage is clearly dominant. It is the leafy green gametophytic stage of the bryophytes that produces the gametes. These swim through moisture, present as a film of either rain or dew, from the antheridia to the archegonia. Here the sperm fertilize the eggs, producing zygotes that make diploid sporophytes by mitotic division. The sporophyte plant is attached to the gametophyte and grows directly from it. The sporophyte produces the **sporangia**, organs that contain and release the asexual spores. Spores are asexual because unlike gametes, they never meet in a sexual union.

The moss gametophyte generation possesses what appear to be, but aren't, true roots, leaves, and stems (see Figure 20.1). Bryophytes have no **vascular tissue**, a critical component in such structures. Rootlike organs in plants without vascular tissues are called **rhizoids**; they function like roots, by anchoring the plant and absorbing water and nutrients. The stem is "stem-like" and sometimes is referred to as the **stalk**. The "leafy" parts are sometimes referred to as **"leaves"** because of a lack of better terminology.

VASCULAR PLANTS

Some of the earliest known vascular plants had **roots** that functioned as holdfasts and absorbed water. They also had vascular tissue for water and nutrient movement. This tissue also provided strength and helped hold the plant up in the air. In addition, these early vascular plants had a **waxy cuticular layer** covering the leaves for water retention. The fossil record indicates there was a trend toward the reduction of the gametophyte generation in favor of the more dominant sporophyte generation, which contained the sporangia. The earliest vascular plants probably produced only one kind of spore from one kind of sporangium, a process called **homospory**. After germination, these spores developed into gametophytes with antheridia and archegonia. They produced the sperm and eggs, respectively. However, the trend toward **heterospory** (the production of two different kinds of spores) is also evident in the overall evolution of the vascular plants.

Following the evolution of the aquatic vertebrates, which took place about 500 million years ago, the first vascular plants to colonize the land appear in the fossil record; at about the same time, some arthropods also began colonizing terrestrial habitats.

A significant innovation unique to the vascular plants is the **seed**, which consists of an embryo and some stored food enclosed within a protective coat. The earliest known fossilized seeds date back 350 million years.

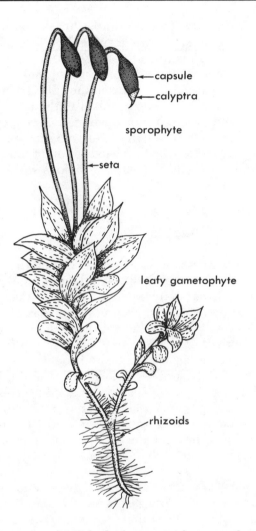

capsule

calyptra

sporophyte

seta

leafy gametophyte

rhizoids

Figure 20.1 *Moss, illustrating rhizoids, leafy gametophytes, and attached sporophytes.*

Five major groups of tracheophytes, or vascular plants, are discussed below. To date, more than 260,000 species of vascular plants have been described.

PSILOPHYTA

The most primitive of the vascular plants are the **psilophytes**. These resemble some of the branching filamentous green algae (Chlorophyta) from which they probably arose. The psilophytes have **true stems**, with simple vascular tissue, branching from slender **rhizomes**, which are elongated, underground, horizontal stems; they are not true roots. The rhizomes have unicellular **rhi-**

zoids, thin, rootlike structures that are similar to root hairs. No true leaves are present, although the aerial stems are green and perform photosynthesis.

Sporangia develop at the ends of the stems and produce haploid spores. These fall to the ground and give rise to the subterranean gametophytes. The gametophytes bear both archegonia and antheridia. Each gametophyte produces both eggs and sperm. The sperm travel to the eggs, where fertilization occurs. Then the diploid zygote begins developing into a sporophyte.

The psilophytes evolved during the Silurian period and thrived more than 300 million years ago. However, with the exception of three surviving species, the entire group is extinct. Many botanists believe some psilopods evolved into the ferns.

LYCOPHYTA: CLUB MOSSES

Lycophyta, or the **club mosses**, also appeared during the Silurian, about 400 million years ago. They were among the largest and most dominant plants during the Devonian and Carboniferous periods. Toward the end of the Permian, they were superseded by more advanced vascular plants. About 1000 species are still found throughout much of the world, although all are quite small and amount to little more than ground cover.

Club mosses have true leaves, stems, and roots. They may have evolved from algae that penetrated the ground, occasionally sending branches above the surface. Some of their leaves are specialized. Called **sporophylls**, these leaves have sporangia which bear spores. Many species have club-shaped structures called strobili at the ends of their stems. The **strobili**, formed from clusters of sporophylls, are the source of the group's common name. It should be pointed out, however, that the club mosses are not related to the true mosses, or bryophytes.

Some club mosses are heterosporous, having two types of sporangia. The larger spores, known as **megaspores**, develop into the archegonia-bearing female gametophytes. The other sporangia produce smaller spores, the **microspores**, which develop into antheridia-bearing male gametophytes.

SPHENOPHYTA: HORSETAILS

Another group of vascular plants is **Sphenophyta**, commonly known as the **horsetails** because that is what they look like. They appeared during the Devonian, around 360 million years ago, and dominated forests during the late Paleozoic era. About 250 million years ago, they began to decline. Today there are some 25 species left, most of which are relatively small.

They all have true roots, stems, and leaves. Living horsetails lack cambium, though many of the larger extinct forms had a **vascular cambium**. This is the layer of cells in the trunk that divides, producing the secondary tissues that allow plants to grow more tissues as they increase in height and weight.

As certain plants grow older, it is the vascular cambium that accounts for their increase in width (girth).

The most common group of living horsetails, *Equisetum*, is homosporous; their spores give rise to gametophytes that bear both archegonia and antheridia.

PTEROPHYTA: FERNS

Many botanists feel that the **pterophytes**, or **ferns**, evolved from the psilophyte stock. Ferns have true roots, stems, and leaves. They first appeared during the Devonian, around 400 million years ago, and became important during the Carboniferous, about 300 million years ago. They then declined, along with the psilophytes, lycophytes, and sphenophytes. However, the decline wasn't nearly so drastic for the ferns as it was for their contemporaneous vascular plants; today there are about 11,000 fern species living throughout the world.

Most forms are quite small, though one group, the tree ferns, is the exception. They have a woody trunk with leaves (fronds) and sometimes attain heights of nearly 25 m (82 ft). This is possible because ferns have well-developed vascular systems containing **xylem** (water-transporting cells) and **phloem** (sugar-transporting cells). But, unlike the conifers and broad-leaved trees, ferns do not have any secondary growth, so it is not possible for the fern trunk to increase in girth. This creates an upper height limit for most ferns and is the reason why their stems often lie on the ground, with only the leaves growing upward.

The dominant stage in the fern's life cycle is the leafy diploid sporophyte (see Figure 20.2). The sporangia are often located on the underside of the frond. Those fronds containing sporangia are called sporophylls. The location of the sporangia on each fern depends on the species.

Most ferns have just one type of spore. Under the proper conditions, all the spores develop into gametophytes that bear both archegonia and antheridia. The gametophytes never develop vascular tissue, and they rarely grow very large. Because their free sperm cells are flagellated and require moisture to reach the egg cells inside the archegonia, ferns tend to be restricted to moist environments. When transplanted, however, they have little trouble growing in drier environments, although fertilization will not occur. Where fertilization does occur, the zygote develops into the sporophyte without passing through a protected seedlike stage.

GYMNOSPERMS

In some classifications, the remaining vascular plants are placed in the group **Spermopsida**. This contains the **gymnosperms** and **angiosperms**. Gymnosperms include cycads, the ginkgoes (just one living member), conifers, and others. The angiosperms include all the grasses, sedges, rushes, and other flowering plants.

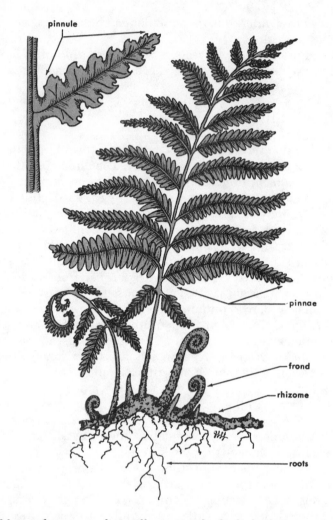

Figure 20.2 *Mature fern sporophyte, illustrating leafy diploid stage.*

It has been found that most of the groups called gymnosperms aren't closely related. Thus, there has been a tendency to list each category separately without placing them under the gymnosperm heading. Here, each of the groups that used to be classified as gymnosperms has been moved up to a taxonomic level that, among botanists, is often called the division. However, the term phylum is more useful.

A distinctive characteristic of the peculiar gymnosperm phylum known as **seed ferns** is not that its species don't produce spores, as true ferns do, but that instead they produce seeds. It is thought that the cycads and angiosperms (flowering plants) evolved from this group.

Another unusual group, the **gnetophytes**, includes about 70 desert and montane (living on mountains) species with flower-like reproductive structures. These plants are heterosporous. The dominant stage in their life cycle is the sporophyte. They are **dioecious** (have separate male and female plants), produce pollen, and bear small, naked seeds. The gnetophytes appear to be relatives of the ginkgo and the conifers. Both the seed ferns and the gnetophytes were important components of Carboniferous forests.

CYCADOPHYTA

The earliest fossils of another ancient group, the **cycads**, date to about 240 million years ago. They reached their climax during the Triassic and Jurassic, about 200 million years ago, and then declined steadily. Now, only about 100 species survive.

The unbranched, erect trunk of the cycad bears a crown of leaves that resembles a palm tree, except for its large, upright **cones**. The tallest cycads reach about 20 m (65 feet). Their leaves are pinnately compound (one vein bearing many leaflets). All the cycads are confined to the tropics, except those grown ornamentally.

Each species has separate male and female plants. The males bear **staminate** cones (with pollen) and the females have seed-bearing, or **ovulate**, cones. There is one exception; instead of bearing cones, the female cycads in the genus *Cycas* have seed-bearing **megasporophylls** that resemble leaves.

Cycad pollen is carried by the wind to the naked ovules. Upon landing, the pollen grain develops a tubular extension called the **pollen tube**. This produces flagellated sperm cells that swim a short distance to reach the egg cells.

GINKGOPHYTA

Before most members became extinct, the group that now consists of only the **ginkgo** was extremely diverse. It was an important constituent of the Mesozoic forests around the world, especially during the Jurassic period, about 150 million years ago. The genus *Ginkgo* is thought to extend all the way back to this period, which is why ginkgoes are often called "living fossils." The group began to decline toward the end of the Mesozoic and continued its decline through the Tertiary until reaching its current situation in which there is just one surviving member, *Ginkgo biloba*. The ginkgo, as it is commonly called, is native to western China, where it has been in cultivation for hundreds of years. During the past hundred years the ginkgo has become a very popular street tree throughout much of the rest of the world.

The fan-shaped leaves are borne on a long stalk (the petiole). The species is dioecious. Male trees have catkin-like **pollen-bearing cones** that, although not very conspicuous, are similar to the male cones found among the conifers. Being heterosporous, the pollen-bearing cones produce **microspores** that are carried by the air. Some land on the female tree's mature ovules. The female gametophytes are buried in the ovule, where the small archegonia are formed. The entrance to the ovule is made through a small opening, the **micropyle**. The microspores are drawn into the micropyle. They lie next to the ovule and

develop into male gametophytes that grow into the ovule. The male gametophyte releases motile sperm cells, one of which fuses with the egg nucleus
to form a zygote, from which the embryo develops. Tissues grow around the
embryo and form a true seed with a soft outer fleshy layer.

CONIFEROPHYTA: CONIFERS

Usually referred to as the **conifers**, this group includes pine, spruce, fir, hem-
lock, cypress, redwood, larch, juniper, and yew, as well as others that aren't
as well-known. The conifers include about 50 genera, comprising over 600
species. Their fossil history dates back to the late Carboniferous, some 300
million years ago. Although the cone is a conspicuous feature distinguishing
many members of this group, it does not appear on all conifers. For example,
juniper berries are actually small cones with fleshy scales. And yews have
seeds surrounded at the base by a fleshy, berry-like pulp. In biological terms,
the distinction between cones and more berry-like forms of reproduction is
significant in terms of the mode of dispersal used by the conifers involved. The
pinoids are adapted for wind dispersal of their seeds. The others are adapted
to animal dispersal.

Trees are the diploid sporophyte stage of the life cycle. The cones are
actually tight clusters of modified leaves known as the sporophylls, which, in
the case of the pines, are also known as **cone scales**. Each sporophyll contains
two sporangia in which haploid spores are produced through meiosis. These
trees are heterosporous. The large female cones contain the sporangia that
produce megaspores; the small male cones contain sporangia that produce
microspores. All seed plants, both the gymnosperms and angiosperms, are
heterosporous.

Each pine scale contains two sporangia; each sporangium has a small
opening, the micropyle. Meiosis occurs inside the sporangium, producing four
haploid megaspores, three of which disintegrate. The remaining megaspore,
through repeated mitotic divisions, becomes the female gametophyte, which,
unlike that in the ferns, is considerably reduced in size. It is located within the
cone. The gametophyte is not free-living, nor does it contain chlorophyll. Each
female gametophyte produces several archegonia in which egg cells develop.
Together the entire structure consisting of the integument, the sporangium,
and the female gametophyte is called an ovule.

Within the male cones are sporangia that produce microspores. These
become **pollen grains**, which develop a thick coating, resistant to desiccation.
They have small wing-like structures that help them along when carried aloft
by the wind. Inside the pollen grains, the haploid nucleus divides mitotically
and the pollen grains become four-celled. Two of these cells degenerate. When
the sporangia burst, millions of mature pollen grains are released. This is the
male gametophyte stage cycle.

Most of the pollen grains that land on the female cones fall between the
scales. Some of these land near the opening of the micropyle. When a pollen
grain lands touching the end of the sporangium inside the micropyle, it grows

a **pollen tube**. The germinated pollen grain, which is now the pollen tube, grows down through the sporangium and penetrates one of the archegonia of the female gametophyte. There the tube releases its nuclei, which, in this case, are sperm that developed when the cells in the pollen tube were dividing. The sperm fertilize the egg and the resulting zygote produces an embryo sporophyte. This is still contained within the female gametophyte inside the pine cone. Finally, the cone sheds its **ovules**, more commonly known as seeds, which grow into adult sporophyte trees.

ANTHOPHYTA/ANGIOSPERMS: FLOWERING PLANTS

The **angiosperms**, or **flowering plants**, also known as the **anthophytes**, are by far the most successful group of living plants, totalling some 250,000 described species. It is their flowers and fruits that differentiate them from all other plants. The earliest known angiosperm-like plants first appear in the fossil record during the Jurassic, about 150 million years ago. But there is some speculation that angiosperms may have existed as long ago as the Permian, about 250 million years ago. It was not until the Cretaceous, 100 million years ago, that there was a rapid decline in the dominance of the gymnosperms. Then, suddenly, and from that time on, the fossil record reflects the diverse spectrum of angiosperms dominating most flora throughout the world.

Among all angiosperms, the diploid sporophyte, retaining and nourishing the gametophyte, dominates the life cycle. And it is the small gametophyte that retains and nourishes the immature eggs and non-motile sperm during development. Flowering plants are heterosporous, having two different sized spores, and oogamous, having sperm and eggs. Their zygotes develop into seeds and fruit that are highly evolved for protection and dispersal. One of their most distinguishing characteristics is the unique reproductive system involving flowers. The floral structures coevolved with the plant's pollinating vectors, which are modes of transferring pollen from one flower to another, such as the wind, insects, and other animals. These vectors appear to have helped angiosperm dispersal into habitats where other plants might not have reached so readily.

Early angiosperms were probably pollinated by the wind. The ovule of modern gymnosperms exudes a sticky substance that traps wind-borne pollen grains. Similarly, the first angiosperms probably had such a pollinating mechanism.

Some insects, such as certain beetles, may have become dependent on readily accessible sticky, sugary droplets, produced by ovules. Insects traveling from one ovule to another may have inadvertently carried pollen with them, conferring a reproductive advantage to some plants, helping to pass on the genes of plants that had larger nectar-secreting organs (**nectaries**) and other structures that lured insects. Plants with certain scents and brightly colored flower parts, arranged in ways that signalled potential pollinators, had selective advantages. These are all insects that specialize in nectar consumption. Because certain insects carried pollen from flower to flower, many became vital to their host plants. Both the plants and the insects benefited from this relationship, and they coevolved. By the early Cenozoic, about 65 million years ago, many modern groups of flowers, as well as the bees, wasps, moths, and butterflies that pollinated them, had already evolved.

TYPICAL FLOWER

The typical flower consists of four whorls of modified leaves called **sepals, petals, stamens**, and **carpels**, all of which are attached to the receptacle. Such flowers are said to be **complete** (see Figure 20.3). Plants with flowers that lack any of the first four elements have **incomplete flowers**.

Usually green because they photosynthesize, the sepals enclose the other parts of the flower inside the bud. Together, the sepals constitute the **calyx**. After the sepals are the petals, the next structures attached to the **receptacle**, which is part of the modified stem. Together, all the petals are called the **corolla**. The calyx and the corolla constitute the **perianth**.

Next, continuing distally (outward) along the modified stem, are the stamens, each consisting of a long, slender stalk, the **filament**, which bears **pollen** on the specialized portion at the tip, known as the **anther**.

The central part of the flower, and the part farthest along the receptacle, is the **carpel**, which is composed of modified floral leaves that have become folded over to protect the ovule from otherwise predaceous pollenators. The **pistil** consists of one or more carpels in the center of the flower. Generally, the pistil has an expanded base, the **ovary**, in which are one or more sporangia. These sporangia are called the ovules. Each one divides meiotically, creating

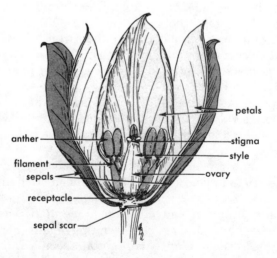

Figure 20.3 Diagram of a tulip flower, illustrating different floral parts.

four haploid megaspores, three of which usually disintegrate. The remaining megaspore divides mitotically to produce the female gametophyte. One of the cells near the micropylar end becomes the egg cell and some polar bodies remain.

Each anther consists of four sporangia. These divide meiotically to produce haploid microspores that mature into pollen grains, each with two nuclei. These pollen grains are the male gametophytes. When the pollen reaches the **stigma**, the receptive portion of the **style** extending above the ovary, a pollen tube grows through the stigma and the style into the ovary. When the tip of the pollen tube reaches the ovule, it enters the micropyle and releases both haploid nuclei into the female gametophyte (embryo sac). One of the haploid nuclei, which are known as sperm nuclei although they are not motile, fertilizes the egg, and the zygote develops into the sporophytic embryo. Two polar nuclei from the female gametophyte combine, forming a diploid fusion nucleus. This in turn combines with the second sperm to form a **triploid** nucleus that divides mitotically and creates triploid tissue known as the **endosperm**. This tissue surrounds the embryo and acts as the stored food upon which the developing sporophytic seed feeds. Endosperm constitutes the nutritious part of most grains and many seeds that are eaten by much of the world's human population.

After fertilization, the angiosperm ovary, which surrounds the embryo, develops into the fruit. This protects the embryo from desiccation during the early stages of development. Later, the fruit acts as an agent of dispersal, either by being blown by the wind or falling to the ground and rolling. An animal may also aid in the dispersal process by carrying the fruit in its fur or ingesting the fruit and depositing it in feces.

The angiosperms are composed of two subgroups, the **Monocotyledonae** and the **Dicotyledonae**. Monocots, as they are often called, have embryos with one **cotyledon**, the leaflike structure composing much of the seed with its endosperm, which is composed of nutrients used during germination. Monocots usually have parallel leaf venation. The flower parts are normally in multiples of three. Most forms lack a vascular cambium. The monocots include grasses, lilies, iris, and orchids. The flowers of grasses have greatly reduced petals, stamens, and carpels, and are thought to have evolved from a lily-like ancestor.

In contrast, the dicots have embryos with two cotyledons, their leaf venation is generally netted, and the lower parts are typically in groups of four or five. In addition, those dicots with secondary growth have a vascular cambium. All the flowering plants not included in the monocots are dicots; they are the most numerous and, in many respects, the most successful plants alive.

Key Terms

algae
alternation of generations
angiosperms
anther

antheridia
Anthophyta
archegonia
brown algae

Bryophyta
calyx
carpels
Chlorophyta
club mosses
complete flowers
cones
cone scales
Coniferophyta
conifers
corolla
cotyledon
Cycadophyta
Dicotyledonae
dioecious
Embryophyta
embryophytes
endosperm
ferns
filament
flowers
gametophyte generation
Ginkgophyta
Gnetophytes
green algae
gymnosperms
heterospory
homospory
hornworts
horsetails
incomplete flowers
leaves
liverworts
Lycophyta
megaspores
megasporophylls
micropyle
microspores
Monocotyledonae
mosses
nectaries
oogamete
oogamous
ovary
ovulate
ovules
perianth

petals
Phaeophyta
phloem
pinoids
pistil
Plantae
pollen-bearing cones
pollen grains
pollen tube
protists
Psilophyta
psilopsids
pterophyta
receptacle
red algae
rhizoids
rhizomes
Rhodophyta
roots
seed
seed ferns
seed plants
sepals
slime molds
Sphenophyta
sporophylls
spermopsida
sporangia
spores
sporophyte generation
stalk
stamens
staminate
stigma
strobili
style
thallus
thallophytes
Tracheophyta
triploid
true fungi
true stems
vascular cambium
vascular plants
vascular tissue
waxy cuticle
xylem

Chapter 20 Self-Test

QUESTIONS TO THINK ABOUT

1. Compare and contrast land plants with those that live in water (give examples).
2. Describe the types of water-proofing used by plants, and what is its function?
3. What are the differences betwen algae and higher plants?
4. What role did vascular tissue play in the evolution of plants?
5. What is a flower? Describe its structure and function.
6. How does alternation of generations apply to different plants?
7. What are the different characteristics that differentiate major groups of plants?
8. How do plants differ from all the other groups of organisms discussed so far?
9. What are seeds? And what is their significance?
10. Describe an angiosperm and give examples of different types of angiosperms.

MULTIPLE-CHOICE QUESTIONS

1. The following is thought to be the group that is the direct common ancestor of most modern plants:

   a. Pyrrophyta (dinoflagellates)
   b. Chlorophyta (green algae)
   c. Phaeophyta (brown algae)
   d. Rhodophyta (red algae)
   e. Euglenophyta (euglenoids)

2. Multicellular algae have

   a. true roots
   b. true stems
   c. true leaves
   d. considerable tissue differentiation
   e. none of the above

3. The simple body form of multicellular algae is called a

   a. thallus
   b. sporangium
   c. micropyle
   d. cone
   e. angiosperm

4. Because the simple body form of multicellular algae is called a thallus, as a group these organisms are sometimes called _____.

    a. sporantiophytes
    b. conifers
    c. angiosperms
    d. algophytes
    e. thallophytes

5. Algal reproductive structures lack

    a. a protective wall of nondividing cells
    b. zygotes that develop into embryos after having been released from the female reproductive organs
    c. a protective wall of sterile cells
    d. all of the above
    e. none of the above

6. Bryophytes include

    a. mosses
    b. liverworts
    c. hornworts
    d. all of the above
    e. none of the above

7. The vascular plants include

    a. club mosses
    b. horsetails
    c. ferns
    d. seed plants
    e. all of the above

8. The following are mechanisms that help embryophytes inhabit terrestrial habitats:

    a. embryophytes have multicellular sex organs covered with a layer of protective, sterile cells.
    b. embryophyte sporangia produce spores that are covered with protective sterile cells
    c. embryophytes have a waxy cuticle that usually covers the aerial parts of the plants
    d. all of the above
    e. none of the above

9. Bryophytes differ from thallophytes (true algae) in that

    a. bryophytes are rarely filamentous, except during one stage in the life history of mosses
    b. bryophytes are composed of cells that form tissues
    c. bryophytes are usually terrestrial, remaining somewhat dependent on their ancestral aquatic environment
    d. all of the above
    e. none of the above

10. Bryophytes
    a. need water for their flagellated sperm cells to swim from the antheridia to the egg cells in the archegonia
    b. lack the ability to move fluids internally across long distances
    c. lack the support of xylem
    d. all of the above
    e. none of the above

11. All bryophytes have a(n)
    a. alternation of generations
    b. sporophyte generation
    c. gametophyte generation
    d. all of the above
    e. none of the above

12. A significant innovation unique to the vascular plants is the seed, which
    a. consists of an embryo
    b. has some stored food
    c. is enclosed within a protective coat
    d. all of the above
    e. none of the above

13. Lycophytes have club-shaped structures at the ends of their stems known as strobili, which form clusters of sporophylls. It is from these structures that the group's common name, _____, has been derived.
    a. strobili mosses
    b. club-shaped mosses
    c. club mosses
    d. sporophyll mosses
    e. cluster mosses

14. Club mosses have
    a. true leaves
    b. true stems
    c. true roots
    d. all of the above
    e. none of the above

15. Horsetails (Sphenophyta) have
    a. true leaves
    b. true stems
    c. true roots
    d. all of the above
    e. none of the above

16. Ferns (Pterophyta) have

    a. true leaves
    b. true stems
    c. true roots
    d. all of the above
    e. none of the above

17. Gymnosperms include

    a. grasses
    b. sedges
    c. rushes
    d. all of the above
    e. none of the above

18. Angiosperms include

    a. cycads
    b. ginkgoes
    c. conifers
    d. all of the above
    e. none of the above

19. The conifers include

    a. pines
    b. spruce
    c. fir
    d. all of the above
    e. none of the above

20. The conifers include

    a. cypress
    b. redwood
    c. larch
    d. all of the above
    e. none of the above

21. The conifers include

    a. hemlock
    b. junipers
    c. yews
    d. all of the above
    e. none of the above

22. The _____ are, in terms of numbers of species, the most successful group of plants.

    a. angiosperms
    b. gymnosperms
    c. ginkgoes
    d. club-mosses
    e. ferns

ANSWERS

1. b	7. e	13. c	18. e
2. e	8. d	14. d	19. d
3. a	9. d	15. d	20. d
4. e	10. d	16. d	21. d
5. d	11. d	17. e	22. a
6. d	12. d		

Animal Kingdom: Animalia

Unicellular organisms are composed of only one cell. So, by definition, they lack tissues and organs, which are aggregations of differentiated cells. Nevertheless, unicellular organisms are extremely complex and often are classified as members of the **animal kingdom**. However, for reasons stated in the previous chapter, these forms have been classified as protists.

Organisms belonging to the animal kingdom share a few fundamental characteristics. When fertilized, or stimulated to divide, the animal egg undergoes several divisions, or **cleavages**, producing a grapelike cluster of cells, the **morula**. As cleavage continues, the increased number of cells becomes arranged in the early embryonic stage known as a blastula, a single spherical layer of cells that encircles a hollow, central cavity. Following the formation of the **blastula**, the cells in this single-layered sphere undergo a series of complex movements establishing the shape and pattern of the early embryo. This transformation process is known as **morphogenesis**. In the process the blastula becomes a simple double-layered embryo, the **gastrula**, from which the organism develops. The adult stages of the animal kingdom are **multicellular**. As adults they are composed of **tissues**, albeit poorly defined tissues in the case of the **sponges (Porifera)**.

All the animal phyla are thought to have evolved from a common ancestor. Well over one million animal species have been classified to date. Of these, the majority by far belong to the phylum **Arthropoda**, most of which are insects (class: **Insecta**). This chapter discusses all of the major phyla and some of the better-known lesser phyla.

LOWER INVERTEBRATES

Porifera

Sponges are a group of about 10,000 species, which are aquatic, mostly marine, multicellular filter feeders that appear to have evolved from protozoans, independently from all the other multicellular phyla. The zygote develops into a multicellular, free-swimming, ciliated larva, which metamorphoses into an adult sponge by turning inside out, bringing its cilia inside. These relatively simple organisms have no organs, and their tissues are not well-defined.

Structurally, they consist of an outer layer of flattened **epidermal cells**. Inside this is a layer of wandering **amoeboid cells** and an innermost layer of **collar cells**, sometimes called **choanocytes**. The collar cells are flagellated. They produce water currents that flow from the external aquatic environment, through small pores in the sponge to the central cavity, the **spongocoel**, and out through a larger opening called the **osculum**. The microscopic food particles brought in with the currents of water are engulfed by the collar cells, which either digest these particles on their own or pass them to the amoeboid cells for digestion. The water currents also deliver oxygen to the cells and carry off carbon dioxide and nitrogenous wastes.

Small needlelike crystals, composed either of **calcium carbonate** or siliceous material (containing **silica**), are scattered throughout the body of the sponge. The crystals, which are called **spicules**, together with **proteinaceous fibers**, create the skeleton that helps these animals maintain their shape.

Sponges' usual modes of reproduction include **asexual budding** as well as **sexual reproduction**; both **monoecious** and **dioecious** forms occur. Although the **larvae** are **ciliated** and free swimming, the **adults** are always **sessile** (sedentary) and are usually attached to a submerged object.

Cnidaria (Coelenterates): Hydra, Jellyfish, Corals, Sea Anemones

The phylum **Cnidaria**, sometimes called **Coelenterata**, includes about 9000 species of jellyfish, hydras, sea anemones, sea fans, and corals. These organisms occur in several basic body forms. Jellyfish are free-swimming adult forms known as **medusae**, which usually look like rounded domes with tentacles hanging below. The **polyp** cnidarian body plan is basically an upside-down medusa. Unlike the motile medusae, polyps have the dome side on the bottom. They are attached to the substrate and therefore are sessile, with their tentacles pointing up, buoyed by the water. **Hydra**, **coral polyps**, and **sea anemones** are examples of the sessile adult cnidarians. Most cnidarians are marine, though there are many freshwater forms.

The entire phylum Cnidaria is composed of simple, radially symmetrical animals composed of an outer tissue layer, the **ectoderm**, and an inner **endoderm**. Between these is a gelatinous filling containing **amoeboid** and **fibrous cells** scattered throughout. This less distinct third tissue layer, the **mesoderm**, in Cnidaria is called the **mesoglea**.

The three tissue layers enclose a hollow interior, which is filled with water that enters and exits through one opening to the outside. By means of simple muscular movement, the jellyfish uses its body as a pump to propel itself through the water.

Cnidarian **tentacles**, attached around the "mouth," have unique structures known as **cnidoblasts**, which are composed of **nematocysts**. These contain specialized harpoon-like structures connected to long threads that are discharged in response to chemical or tactile stimuli. Some of the nematocysts are either sticky, barbed, or poisonous. They have the capacity, in the aggregate, to harpoon, lasso, or paralyze prey. Then the tentacles pull the food to the mouth.

With simple muscle contractions of the hanging tentacles, food is pulled up to the single opening and then into the interior, sometimes called the **gastrovascular cavity**, or **coelenteron**. There it is broken down with enzymes, enabling nutrients to be absorbed by the cells lining the cavity. These then pass the nutrients on to other cells. Waste products pass out the same opening through which the food enters. Beyond the gastrovascular cavity, there are no digestive organs. The nervous system consists of a **nerve net** with no centralization, meaning no central nervous system, and no head.

The only sense organs that any of these animals have are **statocysts**, receptor organs that inform the animal about gravity, and **ocelli**, light-sensitive organs that are groups of pigment cells and photoreceptor cells located at the base of the tentacles. These appear to constitute the first multicellular sense organs.

The hydras (class: **Hydrozoa**) are solitary and have only a polyp stage. However, many other hydrozoans are **colonial** and have a more complex life cycle, with a sedentary hydra-like stage, as well as a free-swimming jellyfish-like medusa stage. Many cnidarians have ciliated, free-swimming larvae known as **planulae**.

The true jellyfish (class: **Scyphozoa**) have a dominant medusa stage, although some also have a planula larva and a polyp stage in the life cycle.

The **sea anemones, sea fans**, and **corals** (class: **Anthozoa**) are all marine. None of the 6,200 members of this class has a medusa stage in the life cycle. They are more complex than the simple hydra-like polyps. The corals secrete a hard, limy skeleton that is a major component of all coral reefs.

Ctenophora: Comb Jellyfish

The **ctenophores**, or **comb jellyfish**, of which only 90 species are known, are very similar to the true jellyfish, being **radially symmetrical** animals with a saclike body composed of an ectoderm, an endoderm, a mesoglea, and a **gastrovascular cavity**. Food is digested by digestive enzymes secreted from the cells lining the gastrovascular cavity. Indigestible material is voided through the mouth and through two small anal pores located near the single statocyst. What also sets ctenophores apart from the cnidarians are their **mesodermal muscles**. In addition, when tentacles are present, they have **adhesive cells** instead of nematocysts. Finally, they lack the polymorphic life cycle found among many of the cnidarians. Most ctenophores have eight rows of cilia; these are the **combs** running along the surface of their transparent body, which account for their ability to swim.

Platyhelminthes: Flatworms, Flukes, and Tapeworms

The 15,000 members of the phylum **Platyhelminthes** are the simplest animals to possess a **bilaterally symmetrical** body plan. This means that they have virtually identical right and left sides, with a different top and bottom (or front and back, depending on how the organism is oriented). They also have head and tail regions. "Headness" is known as **cephalization**, which is typical of most bilateral, active organisms. The head region contains sensory cells, such as nerve cells (**neurons**), and aggregations of nerve cell bodies (**ganglia**), which are considered an early step in the evolution of a brain.

Unlike the cnidarians, which have neurons dispersed in a loose network, termed a nerve net, many of the platyhelminths have nerve cells arranged into long **nerve cords** that carry nerve impulses to and from the ganglia in the anterior end of the body. Many platyhelminths have ocelli that differentiate not only light from dark but also the direction from which the light is coming. Many also form images. In addition, platyhelminths possess **chemoreceptors** that enable them to locate food. Flatworms also have a **tubular excretory system** running the full length of the body. This system contains many small tubules that open at the body surface, where **flame cells** containing **cilia** help move water and waste materials out of the body.

Platyhelminths have three tissue layers—the ectoderm, the mesoderm, and the endoderm—which constitute the cellular portion of the body. All organisms above the ctenophore level of organization have these three distinct tissue layers. Among most **flatworms**, these tissue layers enclose an **internal digestive cavity** that is lined by endoderm. Members of one small group of flatworms, however, known as the **Acoela**, lack a gastrovascular cavity.

In addition to possessing the tissue level of organization, flatworms also have the organ level of complexity. Platyhelminths do not have a cavity, or **coelom**, between their digestive tract and body wall. Therefore, they, together with the **nemertines** are known as **acoelomates**. Nemertines (phylum: Nemertina) are commonly called **ribbon worms** and **proboscis worms** and are characterized by their **proboscis**, a long, muscular, tubular structure capable of being everted from the anterior region.

The platyhelminth classes **Turbellaria (free-living flatworms)** and **Trematoda (flukes)** constitute parasitic flatworms that, unlike the planarians, lack cilia. In contrast to the turbellarians, which have an epidermis, flukes have a thick, highly resistant **cuticle** that protects them from the acids and enzymes of the digestive system(s) in which these parasites often live. Flukes, being internal and external parasites, attach to their animal hosts with characteristic hooks and suckers. In addition to their two-branched gastrovascular cavity, much of their volume consists of reproductive organs. Many flukes have complex life histories, often involving more than one host. Among some forms, the eggs pass through the intestinal cavity of their host and hatch in what is often an aquatic habitat.

The **tapeworms**, class **Cestoda**, are internal parasites, usually living in vertebrate intestines. They have a knoblike "head" structure called the **scolex**, which bears suckers, and many bear hooks as well, enabling them to become attached to their hosts. The long, ribbonlike body is usually divided into segments called **proglottids**, which are little more than reproductive sacs. Each proglottid contains both male and female sex organs. In time, they fill up with mature eggs. When ready, the segment detaches from the worm, moves through the host's intestines, and is passed with the feces. If an appropriate host eats food containing tapeworm eggs, the eggs hatch inside the host. The embryos bore through the intestines, enter the blood, and are carried to the muscles where they **encyst**, becoming encased in a hard protective coating. They remain in the muscles until another animal eats this intermediate host. Inside the new host the cyst's walls dissolve and the young tapeworms develop, attach to the intestinal lining, and continue their life cycle. Intermediate hosts are involved in the life cycles of many tapeworm species.

Cnidarians, ctenophores, and platyhelminths have a gastrovascular cavity with an opening that functions both as a mouth and as an anus. By observing the early embryonic stage, the gastrula, through embryonic development, it has been determined that the **blastopore**, which is the opening connecting the internal cavity, known as the **archenteron**, with the exterior of the gastrula, becomes the common mouth/anus opening in platyhelminths. However, in nemertines, while the blastopore becomes the mouth, an entirely different opening, which appears during embryonic development, becomes the anus. Similarly, it has been found that many other phyla, including the nematode worms, mollusks, annelids, and arthropods, share a similar type of development in which the blastopore becomes the mouth. For this reason, they are often placed together in a group called the **Protostomia**.

Other evolutionary lines in which the blastopore becomes the anus and a separate opening becomes the mouth include such phyla as the echinoderms and chordates. These phyla are collectively termed the **Deuterostomia**.

In addition to the formation of the mouth and anus, there are many other differences between the protostomes and deuterostomes:

1. Cells in early cleavage stages in the protostomes are determinate; that is, the developmental fates of these first few embryonic cells are at least partially determined. The early cleavage stages in the deuterostomes are indeterminate. Because the developmental fates of the first few cells have yet to be determined, each cell can develop into an individual if separated at this early stage.
2. They have different cleavage patterns. That is, they differ in the ways in which the zygote divides into separate cells.
3. Larval types differ.
4. The mesoderm is formed from the ectoderm in what are sometimes referred to as the **radiate phyla** (the cnidarians and ctenophores). Beyond the radiate phyla are the more advanced protostomes and deuterostomes. In the protostomes, only a small amount of the mesoderm is formed from the ectoderm; most of it arises from cells located near the blastopore, between the endoderm and the ectoderm. In the deuterostomes, all of the mesoderm is of endodermal origin.
5. If a coelom is present, its formation differs with regard to the protostomes and deuterostomes. A true coelom is a cavity located between the digestive tract and the body wall, entirely surrounded by mesoderm. Coelomate protostomes have a coelom that forms from a split in the mesoderm. In coelomate deuterostomes the coelom forms from mesodermal sacs. These evaginate (grow out) from the wall of the archenteron, the cavity inside the gastrula stage of the early embryo, and become the digestive tract.

Aschelminthes: Nematodes (Roundworms), Sac Worms, Rotifers

The **aschelminthes** include small, often wormlike animals with a direct, rather straight digestive tract and a protective cuticle. None has a "head," a respiratory system, or a circulatory system. Similar to the platyhelminths, many have a **flame cell excretory system**, although the nematodes have a unique excretory system.

Nematodes, of which there are over 80,000 known species, range in size from microscopic to over 3 feet in length. They are elongate round worms that taper to a point at both ends. Unlike the flatworms, they don't have ciliary movement; instead, they thrash about by alternately contracting their longitudinal muscles. Since the nematode body cavity is not entirely enclosed by mesoderm, in this case represented by bands of muscle, they do not have a true coelom, which is why they are called **pseudocoelomates**.

Rotifers represent about 2,000 species of some of the smallest and most common aquatic organisms. They are named for the circle of cilia at their anterior end. The cilia beat in a manner that appears to be circular, like a rapidly rotating wheel. With the beating cilia, rotifers draw a current of water into their mouths, enabling them to capture unicellular organisms, which are then ground up by hard jawlike structures. A posterior "foot" is used for attachment. Like the nematodes, rotifers are considered pseudocoelomate protostomes.

Lophophorate Phyla

There are a few phyla of small protostomes, all of which have a lophophore, a specialized U- or horseshoe-shaped fold around the mouth with many ciliated tentacles attached. This structure creates a current of water that sweeps unicellular organisms and other tiny particles into a groove that leads to the mouth.

The digestive tracts in many of these organisms are U-shaped, with the anus lying outside the crown of tentacles. All the organisms included in these phyla (**Phoronida, Ectoprocta, Bryozoa**, and **Brachiopoda**) are aquatic and most are marine. Their larvae are ciliated and free-swimming. Eventually they settle down to the substrate, where they secrete a protective case and remain sessile the rest of their lives.

The **phoronids** amount to about 15 wormlike organisms. There are about 5,000 species of **ectoprocts**; most are small, colonial, sessile organisms. The **brachiopods** represent a group of 335 hard-shelled animals that look something like bivalved mollusks (clams). However, the relationships between brachiopods and other invertebrate phyla remain uncertain. Brachiopods were far more common millions of years ago; in fact, over 30,000 extinct species have been described.

HIGHER INVERTEBRATES

Mollusca: Snails, Clams, Octopuses, Chitons

Mollusca and the next phylum, **Annelida**, are both protostomous. It is thought that they both evolved from a segmented common ancestor, which was divided into repeated sections by a series of partitions called **septa**. However, unlike the annelids, the mollusks lost their segmentation. Mollusks and annelids possess a coelom and a **circulatory system**. The coelom divides the muscles of the gut from those of the body wall, enabling both sets of muscles to move independently. Development of the coelom was paralleled by the development of a complex circulatory system, made possible by the coelomic space where fluids bathe the organs without being squeezed out by the surrounding muscles.

The mollusks are one of the more successful animal phyla in terms of species numbers, totalling about 110,000 in all. Most forms have what is termed a **foot**, the muscular organ upon which they move. Between the foot and and the **mantle**, which is the outermost layer of the body wall, are the internal organs. The mantle of most mollusks secretes a calcium-containing **shell**, although some forms such as slugs and octopuses have lost their shells. Others have only reduced, modified versions of what were once more pronounced versions of a shell.

Mollusks share many of the same features. They are bilaterally symmetrical and most have **gills**. Some have an **open circulatory system**, while others have one that is **closed**. Mollusks have considerable cephalization with both **central** and **peripheral nervous systems**. They are dioecious, and their organ systems include specialized structures that possess **nephridia**. Many mollusks have a larval stage that is quite similar to that of marine turbellarians.

Despite the many shared molluscan features, this phylum is also extremely diverse. Although there are seven major groups (classes), the best known are those in **Gastropoda**, which includes the **snails, slugs, nudibranchs, conchs, abalones**, and **whelks**. These organisms have either one shell or have lost it entirely. The class **Bivalvia** includes what are more commonly known as the bivalves, those "shells" with two shells or valves. Among these are the **clams, oysters, scallops**, and **mussels**. The **Cephalopoda** include the **octopuses, squid, cuttlefish**, and nautiluses. The **chitons** are members of the class **Amphineura**. Generally regarded as the most primitive of the living mollusks, their body plan is the closest to what the first mollusks are thought to have looked like. Their peculiar, segmented shell distinguishes them from all other molluscan classes; it consists of eight serially arranged dorsal plates.

Annelids

Annelid worms have ring-like external segments coinciding with internal partitions. The approximately 9,000 species of these worms have been classified into three major groups. The **earthworms**, also called **terrestrial bristle**

worms (class: **Oligochaeta**) are mainly terrestrial and freshwater scavengers that burrow in moist soil. The **marine bristle worms** (class: **Polychaeta**) are marine annelids that typically possess a distinct head with eyes and antennae. Usually, each of the serially arranged body segments has a pair of lateral appendages called **parapodia** for both locomotion and gas exchange. Most polychaetes move about by either swimming or crawling and are often found under things or in the mud and sand. Others are sedentary, living in tubes. Some of the tube dwellers have a crown of colorful, featherlike processes.

Believed to have evolved from oligochaetes, the class **Hirudinea** contains the group commonly called **leeches**. Though sometimes called bloodsuckers, many are not parasitic. They are flattened and possess a sucker at each of the tapered ends. The bloodsucking forms attach themselves with the posterior sucker; with the anterior sucker they either pierce through the host's skin with their sharp jaws or they dissolve the host's flesh with enzymes. Blocking the clotting process by secreting the anticoagulant **hirudin**, they continue to ingest the steady flow of blood.

Most annelids have digestive tracts that are straight, running from the mouth along the entire length of the body to the anus. Some annelids have gills, though these are usually the marine forms with parapodia. Inside the parapodia are capillary beds that function in gas exchange. The blood transport system is closed with hearts located only along the main vessels. Blood is then circulated through many smaller, adjacent vessels. Water and salt regulation is maintained by many kidney-like structures called **nephridia**. Their reproductive systems are well-developed: Some species are dioecious, others monoecious, and some earthworms are known for their **hermaphroditism**, meaning that each individual has both male and female reproductive parts. Less well known is the fact that they still need to mate with one another because their organs are aligned in such a way that they are unable to mate with themselves. Therefore, these earthworms line up together with their reproductive organs meeting in what amounts to a simultaneous double mating.

Asexual reproduction is common among many forms of annelids as well. This is usually accomplished when a parent worm breaks into two or more segments that regenerate. The process is called **fragmentation**. Often, even before the adult divides, **regeneration** precedes the separation, so that a new zone of cells begins to form head and tail parts in the appropriate position. Such zones of regeneration are called **fission zones**.

Onychophora

This small phylum includes only about 80 species worldwide, and all are restricted to tropical regions. Their long wormlike bodies look very caterpillar-like moving about on many pairs of short, unjointed legs. **Onychophorans** have a thin, flexible cuticle much like that of the annelids, rather than a harder, jointed, less permeable, more arthropod-like cuticle. Like the annelids, onychophorans have a pair of nephridia in each body segment, but like the arthropods, they have an open circulatory system. Modern onychophorans also possess a **tracheal respiratory system** that is something like that found among many of the terrestrial arthropods; however, it is thought that these systems evolved independently.

Arthropoda: Crustaceans, Spiders, Mites, Ticks, and Insects

The phylum **Arthropoda** includes more species than any other phylum, with over 500,000 described insect species. Some researchers feel that this may represent only a fraction of their total number, claiming that there might be as many as 10 million living species of insects. It is thought that most species inhabit the tropical rain forests. Insects account for the majority of all the arthropodan species.

In much the same way that the chordates (which include the vertebrates) represent the most successful group of deuterostomes, arthropods are the most diverse and successful group of protostomes.

The tough external cuticle covering arthropodan bodies functions as an external skeleton, or **exoskeleton**. The cuticle is composed of proteins and the strong, but flexible, polysaccharide **chitin**. Attached to the inside of the many jointed, hinged parts, or plates, are muscles that enable the organism to move while being covered with a protective armor. Equally as important is the exoskeleton's value as waterproofing. This has allowed arthropods to become one of the most successful groups to colonize terrestrial habitats.

The main drawback, however, to the arthropodan exoskeleton is that it restricts growth, weight, and size. To grow beyond a certain point, an organism must periodically detach the muscles connected to the interior of the exoskeleton, shed its cuticular layer, expand in size, and then lay down a new hard outer covering to which the muscles are reattached. Growth, therefore, is made possible through a series of molts during an arthropod's lifetime. These successive molts also give the organism an opportunity for morphological change. This is especially true of the young. Larvae of many arthropods go through a succession of genetically controlled changes, thereby incrementally becoming more adult-like. Other species undergo a more rapid transformation, or **metamorphosis**, in just one molt. This is the case with caterpillars which pass through their pupal stage before becoming a butterfly.

The generalized arthropodan body plan is slightly elongate with a segmented body somewhat reminiscent of its annelid-like ancestry. During the course of evolution, many arthropods lost much of their segmentation. In some forms, the segments became grouped together to form distinct body sections.

The open arthropodan circulatory system, with an internal cavity, the **hemocoel**, bathed with its equivalent of blood, **hemolymph**, was described in Chapter 11, on internal transport systems.

One of the major arthropod classes is the **Arachnida (spiders, scorpions, mites**, and **ticks)**. Arachnids have six pairs of jointed appendages, the most anterior of which, the chelicerae, are adapted for manipulating, piercing, and sucking out their prey's fluids. Sometimes poison glands are associated with this first pair of appendages. The next pair, the pedipalps, are sensory. They are sensitive to touch and are capable of detecting certain chemicals; they also help to manipulate food. The other four pairs of appendages are used for locomotion.

Another class in the phylum Arthropoda is **Crustacea (lobsters, shrimp, crabs, crayfish, barnacles, wood lice, pill bugs**, and **water fleas)**. Most forms are aquatic, though some, such as the pill bugs, are terrestrial. They have two pairs of antennae, three pairs of feeding appendages adapted for chewing, and a number of pairs of legs. This is a diverse class in terms of the number and arrangement of their appendages. They are also extremely variable in terms of size: Some are small planktonic forms, others are huge crabs.

The major class in this phylum is **Insecta**. Most insects are terrestrial, and those that are aquatic appear to have evolved from terrestrial forms. Their arthropodan segmentation is reduced to three major body segments, the **head, thorax**, and **abdomen**. The head has a pair of antennae, specialized mouth-parts, and compound eyes. The thorax has three pairs of legs and, often, one or two pairs of wings. The abdomen contains the viscera and reproductive organs. (See Figure 21.1.)

Like other arthropods, insects have specialized excretory **Malpighian tubules** and **secretory glands** that maintain water and salt balance. These structures accomplish functions analogous to those carried out by vertebrate kidneys.

Insects, like other arthropods, are dioecious. Breathing is accomplished through series of pores, the **spiracles**, which carry air through a series of smaller and smaller branching tubes, the **trachea**, to all the cells of the body.

Other arthropod classes include **Chilopoda (centipedes)** and **Diplopoda**, which are commonly called the **millipedes**. Both of these classes have bodies that are divided into a head and trunk region. The latter part is elongate, sometimes somewhat flattened, and divided into many segments. One of the main differences between centipedes and millipedes is that the former have a single pair of legs on each trunk segment while the latter have two pairs of legs per segment, that is, two legs on each side of the body segment.

Echinodermata

The **echinoderms** are one of the two major deuterostomous phyla. Each of the more than 6,000 species of **crinoids, starfish, sea urchins, sea cucumbers,**

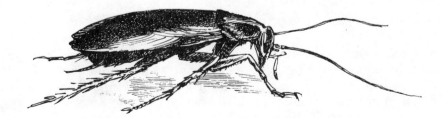

Figure 21.1 Insect: American cockroach.

brittle stars, and **sand dollars** in this group of "spiny-skinned" organisms, the literal meaning of Echinodermata, has a series of **calcareous spines** and **plates** located just under the skin. They also have **pentaradial symmetry**, with adult bodies divided into five parts around a central disc. It is usually underneath this central disc that the mouth is located.

Also unique to the echinoderms is their **water vascular system**, a series of fluid-filled vessels that use hydraulic pressure to operate the hollow tube feet or podia, each of which has a suction tip that can be attached or removed from objects. When coordinated, the tube feet manage feeding and locomotory functions.

All echinoderms are marine, usually bottom-dwelling, and most are motile. The major classes include the **Crinoidea** (sea lillies and feather stars). These have branched, feathery arms that use mucus to trap floating food particles, which are then brought to the mouth by means of the ciliated tube feet.

The **Asteroidea** (sea stars) are what most people call starfish. They are flattened, with five or more arms radiating from the central disc. Most are carnivorous, often feeding on crustaceans and mollusks, which they can open up with sustained pressure applied with their tube feet.

The **Ophiuroidea** (brittle stars and serpent stars) include animals that look something like the sea stars except that they have longer, thinner, more flexible arms branching off from their central disc. The common name comes from their arms, which are very brittle, and snap off when handled. Ophiuroidea eat detritus, organic debris that settles on the bottom, which is then digested in their large, simple, sac-like stomach. They have no intestine or anus.

Echinoidea (sea urchins and **sand dollars**) represent roundish, sometimes almost spherical (e.g., sea urchins), and sometimes flattened organisms (e.g., sand dollars) that live mouth-down on the bottom. Unlike the ophiuroids, they have an intestine and anus in addition to a mouth. Sea urchins are usually protected by spines ranging in size from those which are quite short to some forms with long, thin, pointed spines that can penetrate soft skin. The calcareous plates have holes through which the tube feet protrude.

Holothurioidea (sea cucumbers) look like cucumbers. They don't have any arms but are able to move slowly on their slightly flattened ventral side with their tube feet. Many forms eat sand, filtering out the organic particles and passing the rest through their anus.

Embryological studies have revealed that the echinoderms and the next group, the chordates, share many characteristics. Therefore, researchers place them on an evolutionary branch that diverged long ago from the protostomes.

Hemichordata: Acorn Worms

There is one more invertebrate phylum, **Hemichordata**, to describe before reaching the chordates. This represents about 85 marine animals that look something like fairly large worms, most of which live in U-shaped burrows. Their conical proboscises are acorn-like—hence the name, acorn worms.

They are particularly significant because they share some characteristics of both the echinoderms and the chordates. Therefore, they are sometimes thought of as an offshoot from an early common ancestor. Along the wall of their pharynx is a series of **gill slits**, one of the key characteristics that identify the chordates. In addition, they have a ciliated larval stage that is very similar to that of some echinoderms.

CHORDATA

Urochordata: Tunicates and Sea Squirts and Cephalochordata: Lancelets

The phylum **Chordata** is usually divided into three subphyla, the **Urochordata (tunicates)**, **Cephalochordata** (lancelets), and the **Vertebrata** (vertebrates). The first two subphyla are considered the invertebrate chordates, although each contains, during some stage in its life history, a rod-like **notochord** that acts as an internal skeleton. In the vertebrates, it is the notochord that is surrounded or replaced by vertebrae. All three subphyla have gill slits in their pharynx (pharyngeal gill slits) at some point during their lives. They also have a **dorsal hollow nerve cord**, unlike other animals with a main nerve cord that is ventrally located.

   Urochordata (tunicates and sea squirts) have tadpole-like larvae with a notochord and a dorsal hollow nerve cord. During metamorphosis, the notochord is resorbed and the larva expands in girth into a sessile adult tunicate, which is little more than a sedentary, filter-feeding pharyngeal basket.

   Only 29 species in the subphylum **Cephalochordata** are known. Although similar to the urochordate tadpole larvae just described, the lancelets (**Amphioxus**) have a much larger gill system containing many more gill slits. They feed like urochordates, using cilia to create a water current that carries particulate matter into their mouths. While the water passes out the gill slits, the food particles trapped in mucus pass down the pharynx into the alimentary canal where they are digested.

Vertebrata

Vertebrates (**Vertebrata**) have a vertebral column that distinguishes them from other chordates. Their vertebrae form the backbone that is the supporting axis holding up the body and protecting the spinal cord. Together, the seven classes composing this group total about 43,000 species, of which nearly half are **fish** (superclass: **Pisces**). The fish comprise three classes, **Agnatha** (**lampreys** and **hagfish**), **Chondrichthyes** (**sharks** and **rays**), and **Osteichthyes** (**boney fish**). The other four classes are the **amphibians, reptiles, birds**, and **mammals**.

Pisces: Fish

Agnathans were originally filter-feeders, straining mud and water through their mouth and out their gills. The only living members are highly modified, having lost their bone and replaced it with cartilage. Both lampreys and hagfish are

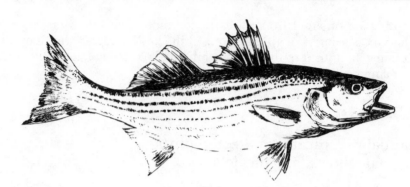

Figure 21.2 Fish: Striped bass.

jawless, with disc-shaped mouths that are either rasping or sucking. Their gills are internal to their gill arches. They lack scales and, unlike the other two classes of fish, they lack paired fins. Lamprey larvae are filter-feeding and remarkably similar to *Amphioxus*.

Chondrichthyes include the sharks, skates, rays, and chimaeras. They all have cartilaginous skeletons, paired fins, and jaws derived from gill arches. Living forms have teeth and small scales.

All the other fish are members of the group **Osteichthyes**. Most are boney and all have jaws. (See Figure 21.2.) An early offshoot of this group eventually led to the first amphibians.

Amphibia: Amphibians

The earliest **amphibians** were quite fishlike. Although they had lungs and leg-like appendages, they probably spent most of their time in the water. Slowly they exploited nearby terrestrial habitats, but specific characteristics restricted their advancement, evolutionarily speaking.

They all laid fish-like eggs that were generally exposed and susceptible to rapid desiccation. Therefore, these eggs needed to be laid in a moist place and were usually deposited in the water. From behavioral studies of modern amphibians, it is thought that most early amphibians were external fertilizers; that is, sperm was deposited and fertilization occurred after the eggs had been laid. This had the effect of restricting the breeding of amphibians to times when the males and females could meet in the same spot where the eggs are laid. Most modern amphibians are external fertilizers, though some have evolved modes of internal fertilization.

During development, the norm is for the young to pass through a gilled larval stage, although in a very few cases, the larval stage has been circumvented through specialized developmental modifications. Most adult amphibians have a rather **permeable skin** that renders them susceptible to desiccation. Amphibians played a dominant role in life on earth during the Carboniferous period, often called the Age of Amphibians, 280 to 360 million years ago.

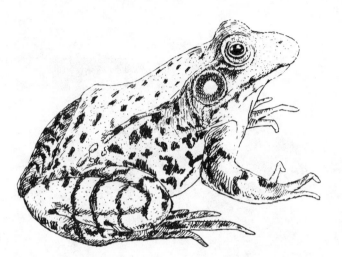

Figure 21.3 *Amphibian: Green frog.*

However, they slowly declined as members of a new class of chordates, the reptiles, replaced them.

Of those amphibians to survive, the groups that are still represented include the salamanders (order: **Apoda**), and the frogs (order: **Anura**). Together these living amphibian species total about 2,500, and new species, especially frogs, continue to be discovered. (See Figure 21.3.)

Reptilia (Reptiles)

Evolving early on from primitive amphibians, the reptiles expanded in numbers and importance until the Mesozoic era, or the Age of Reptiles, 255–65 million years ago, after which their dominance declined. The characteris-

Figure 21.4 *Reptile: Garter snake.*

Figure 21.5 *Reptile: Snapping turtle.*

tics that enabled them to become so successful included both the **shelled egg** and the embryonic membrane, known as the **amnion**, which the amphibians lacked. The amnion is the innermost membrane in the reptilian egg. It is also found in bird eggs as well as among mammals. The embryo, located within the amnion, is bathed in amniotic fluid, which protects the developing animal from mechanical injury. It seems to have been an important development that, in addition to the outer shell, enabled reptiles to lay terrestrial eggs, freeing them from their amphibian and fish ancestry, which required aquatic egg-laying.

Reptiles are **internally fertilized**. The male inseminates the female with sperm. Then the female can go off and lay her eggs at another time. In some instances, the sperm can live over a year inside the female, giving her considerable freedom with regard to where and when she oviposits (lays her eggs). When the eggs hatch, unlike the amphibians, the young are basically small versions of the adults, never having to pass through a larval stage. Since amphibian larvae are tied to the water, this advancement created another factor that freed reptiles from their aquatic ancestry.

Some reptiles were largely terrestrial; their dry, scaly, relatively impermeable skin substantially reduced their susceptibility to water loss, allowing them to inhabit environments that were considerably more arid than those occupied by most other vertebrates.

Other differences include the reptilian legs, which are usually larger and stronger and are oriented less laterally than those of the amphibians, enabling many reptiles to carry their bodies off the ground. This lateral arrangement appears to have been better for more rapid, terrestrial locomotion. Reptilian

Figure 21.6 Bird: Black-crowned night heron.

lungs are better developed than those of amphibians, and greater rib musculature enables better lung ventilation. Their four-chambered heart is better equipped to cope with the demands of a terrestrial existence.

The four surviving orders of modern reptiles are the turtles (order: **Chelonia**), crocodiles and alligators (order: **Crocodilia**), lizards and snakes (order: **Squamata**), and the tuatara, which is the sole surviving member of an ancient order, **Rhynchocephalia**. Together, there are about 6,500 species of reptiles. (See Figures 21.4 and 21.5.)

Aves: Birds

The earliest bird fossils are those of **Archaeopteryx**, which appear in 150 million year old deposits. These animals had wings and feathers. But, unlike modern birds, *Archaeopteryx* had teeth and a long boney tail. Modern birds lack teeth, and their bones are hollow, making them lighter, an adaptation for flight. Like reptiles, they have a four-chambered heart. There are nearly 9,000 species. All have internal fertilization and land-adapted eggs with calcium carbonate shells. They are **homeothermic**. That is, they regulate their body temperature, in large part by means of their metabolism. (See Figure 21.6.)

Mammalia: Mammals

Mammals also evolved from an early group of reptiles. The oldest mammalian fossils are 200 million years old, from the Triassic. Reptiles were still domi-

Figure 21.7 Mammal: Eastern chipmunk.

nating the Earth when the first mammals evolved, but it was during this time that mammals established many of the characteristics that eventually enabled them to do so well when the opportunities became available. (See Figure 21.7.)

It was at the end of the Cretaceous period, about 65 million years ago, that many of the dominant groups of reptiles became extinct over a span of several million years. This period of extinctions was accompanied by increased opportunities that led to an increase in mammalian numbers as well as species.

Like the birds, mammals have a four-chambered heart. Both mammals and birds are homeothermic, or warm-blooded. Both bird feathers and mammal **hair** and **fur** function as insulation. Other characteristics that differentiate the mammals from the reptiles include their **diaphragm**, the muscle under the rib cage that significantly improves their breathing efficiency. Specific skeletal features are unique as well as certain anatomical features, such as their greatly enlarged cerebrum. Mammals do not lay eggs, except for the monotremes, which include the platypus and the spiny anteater. After birth, mammals nourish their young with the milk that is secreted from the mother's **mammary glands**.

Key Terms

abalones	animal kingdom
abdomen	Annelida
Aceola	annelid worms
acoelomates	Anthozoa
acorn worms	Anura
adhesive cells	Apoda
Agnatha	Arachnida
alligators	archenteron
amoeboid cells	Arthropoda
amphibians	Aschelminthes
Amphineura	asexual budding
Amphioxus	Asteroidea

amnion
Aves
barnacles
bilateral symmetry
birds
Bivalvia
blastopore
boney fish
Brachiopoda
brittle stars
Bryozoa
calcareous spines and plates
calcium carbonate
centipedes
central nervous system
cephalization
Cephalochordata
Cephalopoda
Cestoda
Chelonia
chemoreceptors
Chilopoda
chimaeras
chitin
chitons
choanocytes
chondrichthyes
Chordata
cilia
ciliated larvae
clams
cleavage
Cnidaria
cnidoblast
coelenterates
coelenteron
coelom
collar cells
colonial
comb jellyfish
combs
conchs
corals
crabs
crayfish
crinoids
crocodiles

Crocodilia
Crustacea
crustaceans
Ctenphora
cuticle
cuttlefish
Deuterostomia
diaphragm
dioecious
Diplopoda
dorsal hollow nerve cord
earthworms
Echinodermata
Echinoidea
ectoderm
Ectoprocta
encyst
endoderm
epidermal cells
excretory system
exoskeleton
feathers
fibrous cells
fish
fission zones
flame cells
flatworms
flukes
foot
fragmentation
frogs
fur
ganglia
Gastropoda
gastrovascular cavity
gastrula
gills
gill slits
hagfish
hair
head
Hemichordata
hemocoel
hemolymph
hermaphroditism
hirudin
Hirudinea

Holothurioidea
homeothermic
Hydra
Hydrozoa
Insecta
insects
internal digestive cavity
internally fertilized
jellyfish
lampreys
lancelets
leeches
lizards
lobsters
lophophorates
Malpighian tubules
mammals
mammary glands
mantle
marine bristle worms
medusa
mesoderm
mesodermal muscles
mesoglea
metamorphosis
millipedes
mites
Mollusca
monoecious
morphogenesis
morula
multicellular
mussels
nautiluses
nematocyst
nematodes
nemertines
nephridia
nerve cords
nerve net
neurons
notochord
nudibranchs

ocelli
octopuses
Oligochaeta
Onychophora
open circulatory system
Ophiuroidea
osculum
Osteichthyes
oysters
parapodia
pentaradial symmetry
peripheral nervous system
permeable skin
Phoronida
pill bugs
Pisces
planula
Platyhelminthes
Polychaeta
polyp
Porifera
proboscis worms
proglottids
proteinaceous fibers
Protostomia
pseudocoelomates
radial symmetry
radiate phyla
rays
regeneration
reptiles
Rhynchocephalia
ribbon worms
rotifers
roundworms
sacworms
salamanders
sand dollars
scallops
scolex
scorpions
Scyphozoa
sea anemones

sea cucumbers
sea fan
sea squirts
sea urchins
secretary glands
septa
sessile adults
sexual reproduction
sharks
shell
shelled egg
shrimp
silica
slugs
snails
snakes
spicules
spiders
spiracles
sponges
spongocoel

Squamata
squid
starfish
statocysts
tapeworms
terrestrial bristle worms
thorax
ticks
tissues
trachea
trachea respiratory system
Trematoda
tunicates
Turbellaria
turtles
Urochordata
vertebrates
water fleas
water vascular system
whelks
wood lice

Chapter 21 Self-Test

QUESTIONS TO THINK ABOUT

1. What features do members of the animal kingdom share?
2. What are all the different kingdoms of organisms and what features characterize each one?
3. What differentiates members of the animal kingdom from the other kingdoms?
4. List as many phyla within the kingdom Animalia as you can think of, and then write down their main characteristics.
5. What differentiates the protostomes from the deuterostomes?
6. Are humans more closely related to starfish or to insects? Why?
7. Compare and contrast the arthropods and the mollusks. Include major groups of each phylum.
8. What features do all chordates have in common? And do they share any of these with a possible common ancestor? If so, what is that common ancestor thought to be? Describe a living relative.
9. What characteristics are used to classify the major groups (classes) of vertebrates? Use those characteristics to describe each of the vertebrate classes.

MULTIPLE-CHOICE QUESTIONS

Sponges

1. Sponges are multicellular filter-feeders that have _____ organs.

   a. no
   b. one
   c. two
   d. three
   e. four

2. The poorly defined tissue layers of poriferans include

   a. an outer layer of flattened epidermal cells
   b. a layer of wandering amoeboid cells
   c. an innermost layer of collar cells
   d. all of the above
   e. none of the above

3. Sponges have a central cavity called the _____.

   a. osculum
   b. spicule
   c. spongocoel
   d. gastrovascular cavity
   e. stomach

4. Water leaves the spongocoel via the _____.

   a. osculum
   b. spicule
   c. spongocoel
   d. gastrovascular cavity
   e. stomach

5. The water currents also deliver oxygen to the cells and carry off _____ and _____.

   a. carbon dioxide and nitrogenous wastes
   b. chlorophyll and carotenoids
   c. leucocytes and lymphocytes
   d. all of the above
   e. none of the above

6. Small needlelike crystals of either calcium carbonate or siliceous material, known as _____, are scattered throughout the body of the sponge creating a skeleton of sorts that helps maintain their shape.

   a. spongocoels
   b. osculums
   c. spicules
   d. epidermal nodules
   e. mesodermal nodules

7. Adult sponges are _____.

    a. mobile
    b. ciliated
    c. sessile
    d. all of the above
    e. none of the above

Cnidaria and Ctenophora

8. The free-swimming form found among cnidarians that usually looks like a rounded dome with hanging tentacles is known as a _____.

    a. polyp
    b. medusa
    c. snail
    d. mesoglea
    e. planula

9. Cnidarians have a sessile stage that is attached to the substrate, with the tentacles pointing up, which is known as a _____.

    a. polyp
    b. medusa
    c. snail
    d. mesoglea
    e. planula

10. The outer layer of a cnidarian is composed of the tissue known as _____.

    a. mesoderm
    b. phisoderm
    c. endoderm
    d. placoderm
    e. ectoderm

11. Between the cnidarian's outer and inner layers is a gelatinous filling comprising a third layer, containing amoeboid and fibrous cells scattered throughout; this layer is called the _____.

    a. mesoglea
    b. phisoderm
    c. endoderm
    d. placoderm
    e. ectoderm

12. Before the food is absorbed by the cnidarians' cells, it is broken down in the hollow interior called the _____.

    a. gastrovascular cavity
    b. coelenteron
    c. spongocoel
    d. all of the above
    e. a and b

13. Unique to the cnidarians, their tentacles are armed with _____ containing _____ which are like harpoons connected to long threads that are discharged in response to chemical or tactile stimuli.

    a. statocysts, ocelli
    b. nematocysts, cnidoblasts
    c. statocysts, nematocysts
    d. nematocysts, ocelli
    e. cnidoblasts, nematocysts

14. The only sense organs that any of the cnidarians have are _____ and _____.

    a. statocysts, ocelli
    b. cnidoblasts, nematocysts
    c. statocysts, statoblasts
    d. planularians, hydrolarians
    e. scyphozoa, anthozoa

15. Many cnidarians have a ciliated, free-swimming larva known as a _____.

    a. medusa
    b. polyp
    c. planula
    d. tadpole
    e. hydra

16. The _____ are very similar to the true jellyfish, being radially symmetrical animals with a saclike body form composed of an outer ectoderm and an inner endoderm with a mesoglea and gastrovascular cavity. However, they don't have nematocysts, and most have eight rows of cilia (combs) running along the surface of their transparent body, enabling them to swim.

    a. Scyphozoa
    b. Anthozoa
    c. Hydrozoa
    d. Ctenophora
    e. Cnidaria

Platyhelminthes, Nemertea

17. Flatworms have a tubular excretory system running the full length of the body. This system contains many small tubules that open at the body surface where _____ that contain cilia help move water and waste materials out of the body.

    a. chemoreceptors
    b. nerve net
    c. ganglia
    d. kidneys
    e. flame cells

18. Platyhelminths, along with another similar phylum, the Nemertea, are termed _____ because they do not have a cavity between their digestive tract and body wall.

   a. coelomates
   b. acoelomates
   c. protostomes
   d. deuterostomes
   e. proboscis worms

19. Tapeworms have a knoblike "head" structure termed the _____ _____ bearing suckers and often hooks as well.

   a. ganglion
   b. proglottid
   c. scolex
   d. osculum
   e. spicule

20. Tapeworms, members of the platyhelminth class, Cestoda, have a very long ribbonlike body that is usually divided into segments called _____ _____.

   a. ganglia
   b. metamers
   c. septa
   d. scolexes
   e. proglottids

21. Both the cnidarians and the platyhelminths have a gastrovascular cavity with one opening that functions as both a mouth and an anus. During embryonic development, it is the _____ that becomes the mouth and anus in these organisms.

   a. gastropore
   b. archenteron
   c. protostomia
   d. deuterostomia
   e. blastopore

22. It has been shown that in nemertines (proboscis worms) the blastopore becomes the mouth and an entirely different opening, which appears during embryonic development, becomes the anus. This evolutionary line, where the blastopore becomes the mouth and another opening becomes the anus is often called the _____.

   a. lower invertebrates
   b. higher invertebrates
   c. rhynchocephalia
   d. protostomia
   e. deuterostomia

23. Another evolutionary line in which the blastopore becomes the anus and a new mouth is formed, is termed the _____.

   a. cleavagestomia
   b. blastopores
   c. protostomia
   d. deuterostomia
   e. cnidaria

Aschelminthes, Lophophores

24. Nematodes are long round worms that taper to a point at both ends. Unlike the flatworms they don't have ciliary movement, but, instead, they thrash about with their longitudinal muscles. Since their body cavity is not entirely enclosed by mesoderm, they do not have a true coelom, and are termed _____.

   a. acoelomates
   b. coelomates
   c. pseudocoelomates
   d. gastrocoels
   e. enterocoels

25. There are a few phyla of small protostomes, all having a _____, which is a specialized fold around the mouth, usually U-shaped, that has many ciliated tentacles attached. These groups are often called the lophophorates.

   a. tentaculophores
   b. phoronida
   c. bryozoan
   d. brachiopod
   e. lophophore

Mollusca, Annelida, Onychophora

26. The phylum that includes the gastropods, bivalves, cephalopods, and amphineurans is termed _____.

   a. Gastropoda
   .b. Bivalvia
   c. Amphibeura
   d. Cephalopoda
   e. Mollusca

27. The phylum including the earthworms, polychaetes, and leeches is called _____.

   a. Oligochaeta
   b. Polychaeta
   c. Hirudinea
   d. Annelida
   e. Nematoda

28. Earthworms maintain water and salt regulation with many kidney-like structures called the _____.
   a. flame cells
   b. metanephros
   c. nephridia
   d. trachea
   e. ureters

29. Crustaceans, spiders, mites, ticks, and insects are all members of the phylum _____.
   a. Echinodermata
   b. Onychophora
   c. Arthropoda
   d. Arachnida
   e. Insecta

30. Lobsters, shrimp, crabs, crayfish, barnacles, wood lice, pill bugs, and water fleas are all members of the following class of organisms: ____ _____.
   a. Echinodermata
   b. Onychophora
   c. Crustacea
   d. Chilopoda
   e. Diplopoda

31. Arthropods have specialized excretory structures that accomplish functions analogous to those carried out by the kidneys in the vertebrates. These arthropod structures are called _____.
   a. kidneys
   b. spiracles
   c. Malpighian tubules
   d. metanephros
   e. flame cells

32. Insects have tiny pores called _____ that allow air to pass from outside their cuticle to the interior parts of their body.
   a. spiracles
   b. Malpighian tubules
   c. spicules
   d. blastopores
   e. gastropores

33. The centipedes are members of the arthropod class _____.
   a. Chilopoda
   b. Diplopoda
   c. Crustacea
   d. Bivalvia
   e. Onychophora

34. The millipedes are members of the arthropod class _____.

    a. Chilopoda
    b. Diplopoda
    c. Crustacea
    d. Bivalvia
    e. Onychophora

35. Which of the following groups of organisms has pentaradial symmetry? _____.

    a. crinoids
    b. starfish
    c. sea urchins
    d. sea cucumbers
    e. all of the above

36. Without the following, it wouldn't be possible for the tube feet or podia, which enable echinoderms to move about, to function: _____.

    a. water vascular system
    b. hydraulic pressure
    c. fluid-filled vessels
    d. all of the above
    e. none of the above

37. Sea lillies and feather stars have branched, feathery arms that use their mucus to trap food particles floating by. These forms are members of which echinoderm class? _____.

    a. Crinoidea
    b. Asteroidea
    c. Ophiuroidea
    d. Echinoidea
    e. Holothurioidea

38. Starfish, or sea stars, are members of which echinoderm class? _____.

    a. Crinoidea
    b. Asteroidea
    c. Ophiuroidea
    d. Echinoidea
    e. Holothurioidea

39. A marine phylum that contains about 85 species that possess a series of gill slits along the wall of their pharynx, as well as having ciliated larvae, is known as _____.

    a. Echinodermata
    b. Chordata
    c. Hemichordata
    d. Annelida
    e. Arthropoda

Chordata

40. The organisms that at some stage in their life history contain a noto-cord, a dorsal hollow nerve cord, and pharyngeal gill slits are known as

_____.

a. Urochordata
b. Cephalochordata
c. Vertebrata
d. Chordata
e. all of the above

41. The tunicates and sea squirts are members of one of the chordate subphyla known as _____.

a. Urochordata
b. Cephalochordata
c. Hemichordata
d. Vertebrata
e. Chordata

42. Unlike the _____ and the _____, the vertebrates have a vertebral column.

a. Urochordata, Cephalochordata
b. Agnatha, Chondrichthyes
c. Osteichthyes, Aves
d. lampreys and hagfish
e. all of the above

43. What group of chordates that are evolutionarily more advanced than the fish have fish-like eggs and are external fertilizers? _____.

a. amphibians
b. reptiles
c. birds
d. mammals
e. ostracoderms

44. The characteristics that enabled the following group to become so suc-cessful during the Mesozoic era included their shelled eggs and internal fertilization.

a. amphibians
b. reptiles
c. birds
d. mammals
e. fish

45. Unlike modern birds, _____ had teeth and a long boney tail, but they also had feathers.

    a. *Archaeopteryx*
    b. early amphibians
    c. early reptiles
    d. early mammals
    e. none of the above

46. Besides modern birds, the following group of chordates also has a four-chambered heart:

    a. fish
    b. amphibians
    c. reptiles
    d. mammals
    e. hemichordates

47. Like the birds, mammals are _____.

    a. homeothermic
    b. poikilothermic
    c. amniotes
    d. a and b
    e. a and c

48. After birth, mammals nourish their young with milk that is secreted from the _____ of the mother.

    a. hair
    b. mammary glands
    c. sebaceous glands
    d. salivary glands
    e. none of the above

ANSWERS

1. a	13. e	25. e	37. a
2. d	14. a	26. e	38. b
3. c	15. c	27. d	39. c
4. a	16. d	28. c	40. e
5. a	17. e	29. c	41. a
6. c	18. b	30. c	42. a
7. c	19. c	31. c	43. a
8. b	20. d	32. a	44. b
9. a	21. e	33. a	45. a
10. e	22. d	34. b	46. d
11. a	23. d	35. e	47. e
12. e	24. c	36. d	48. b

Index